WITHDRAWN
LIBRARY
College of St. Scholastica
Duluth, Minnesota 55811

MANAGEMENT APPLICATIONS OF DECISION THEORY

MANAGEMENT APPLICATIONS OF DECISION THEORY

JOSEPH W. NEWMAN

The University of Michigan

HARPER & ROW, PUBLISHERS

New York   Evanston   San Francisco   London

658.403
N55

The material in Chapters 3, 6, and 8 is taken from teaching cases written by the author, Copyright © 1964, 1965 by the Board of Trustees of the Leland Stanford Junior University. Used by permission. This material was first published in Joseph W. Newman, *Marketing Management and Information* (Homewood, Ill.: Richard D. Irwin, Inc., 1967).

MANAGEMENT APPLICATIONS OF DECISION THEORY

Copyright © 1971 by Joseph W. Newman

Printed in the United States of America. All rights reserved. No part of this book may be used or reproduced in any manner whatsoever without written permission except in the case of brief quotations embodied in critical articles and reviews. For information address Harper & Row, Publishers, Inc., 49 East 33rd Street, New York, N.Y. 10016.

Standard Book Number: 06-044799-0

Library of Congress Catalog Card Number: 78-148453

CONTENTS

74194

LIBRARY
College of St. Scholastica
Duluth, Minnesota 55811

PREFACE

This book was written for the student of management, in school and out, who would like to find out what Bayesian decision theory is and what is involved in applying it. The focus is on practical applications under real-life circumstances, a feature that distinguishes the book from most writings on the subject.

Business use of decision theory is young but growing. The reason for the increasing interest is that the approach promises to enhance management's ability to make good choices under uncertainty. The book acquaints the reader with the reasons for this promise and gives him a basis for deciding for himself whether the approach will work for him. Major emphasis is placed on the making of requisite managerial judgments as well as on the techniques for using them. The greatest benefit from the use of Bayesian analysis may well be its effect on the way executives think about their problems.

The book is organized in two parts. Part I consists of two introductory chapters on the approach. They presuppose no quantitative background on the part of the reader other than basic algebra. Part II contains nine chapters that describe applications of decision theory made both with and without the computer. These chapters are grouped into three sections. The first two sections describe business situations of major importance, then subject them to detailed analysis. The material is presented in a way that allows the reader to assume the role of decision maker. The contexts are described in sufficient detail to provide him with an opportunity to develop his own approaches to the problems. Study

Questions are given for his guidance. Full commentaries on aspects of the applications follow in later chapters. They take the reader step by step through the development of the author's approaches and discuss related operational problems—after the reader has had the chance to work through those steps by himself. Section Three consists of concluding observations and comments.

THE DECISION SITUATIONS. The Bayesian approach can be applied to any choice problem. Different kinds of decisions are represented in this book. The primary reason for the selection of the subject decision problems was their suitability for exploring the main aspects of decision-theory application under realistic conditions of problem complexity and management operating pressures. Hence, they provide experience in making choices under uncertainty that can be helpful in approaching topically different problems faced by firms offering consumer or industrial products or services.

The issue posed in Part II, Section One, is whether to make a major packaging change that would involve higher costs and, perhaps, an increase in price. The problem was faced by the executives in charge of Maxwell House Coffee. In the handling of the problem described in this book, attention is given to anticipated effects on consumer demand and reactions of competitors. The problem provides the opportunity to use decision trees for structuring both prior and preposterior analyses. In regard to the latter, a number of questions that have not received much attention in the business literature are considered. Of central concern is the question of whether an attempt should be made to obtain additional information before making the final decision and, if so, which of several alternative research plans should be followed. Attention is given to the handling of the imperfect, heterogeneous information that comes from market tests and consumer surveys so that the latter might have appropriate influence on the decision.

A more complex decision problem is the subject of Section Two. It arose when evidence indicated that the General Foods Corporation was about to achieve a major technical innovation—the successful application of the freeze-dry process for producing a soluble coffee. In taste tests on laboratory-produced samples, consumers rated the new product as equal in flavor to ground coffee. A soluble coffee that would be so rated had long been the objective of research and developmental efforts. With the apparent breakthrough, management had to choose among alternative paths for technical development that had different probabilities of success. A central issue was how the new product, which entailed substantially higher costs than the soluble coffees already on the market, should be positioned in the company's multibrand coffee line. This involved considerations of market segmentation, pricing, consumer demand, cannibalism of sales of the company's existing brands, branding, the sequencing of possible freeze-dried-coffee brands, the expected nature and timing of future freeze-dried-coffee competition, and investment.

The freeze-dried-coffee case is especially valuable for providing experience in identifying alternative courses of action, reducing the number to a manageable size, and structuring a complicated problem so that it can be analyzed in an orderly manner. Handling of the case involves the development of a model for the analysis, identification and procurement of the values needed, and utilization of the computer.

SUGGESTED USES OF THE BOOK. This book was written for both classroom use and independent study by business managers and researchers. It is suitable for supplemental use in courses in decision making, research, the use of information, and the application of quantitative methods. Instructors now using other materials on quantitative concepts and techniques may wish to use this book as a supplement to give more attention to the making of business applications of Bayesian analysis. The material has been used successfully in executive development programs.

The book was organized to provide for flexibility in its use, recognizing that instructors will wish to employ it in various ways and allocate various amounts of time to it. The variations will depend on the nature and the level of the course, student backgrounds, interests and teaching techniques of the instructor, and availability of computer facilities for classroom purposes. The book can be read and discussed as a descriptive text. Presentation of the material was planned, however, so that it could also be taught by the case-problem method of instruction, in which the students, aided by Study Questions, work on the problems posed in the chapters before going on to read and discuss the author's commentaries.

Familiarity with the main features of the Bayesian approach can be obtained readily by reading Chapters 1 and 2. A reasonable understanding of what is involved in applications of Bayesian prior and preposterior analysis can be gained from reading and discussing the three chapters that comprise Part II, Section One in about three days. More intensive work, suggested by the Study Questions, may take several days longer. Those who plan to devote no more than half a dozen class periods to the material may find it advisable to confine their work to the first five chapters that precede Part II, Section Two. Time devoted to Section Two can vary from several days for a reading of the chapters to an entire academic term for reading and working intensively on formulating and screening alternatives, modeling, development of input data, and the use of the computer for sensitivity analysis.

Although use of the computer can be advantageous, it is not requisite to realizing most of the book's potential value. For those who wish to make use of the computer, three programs and related materials are

I was fortunate to have the expert secretarial assistance of Miss Gloria E. Bonilla during the time much of the manuscript was being prepared.

The conclusions, opinions, and other statements made in this book are those of the author and are not necessarily those of the organizations or individuals who supported my project or assisted me in my work.

JOSEPH W. NEWMAN

PART I

INTRODUCTION TO THE BAYESIAN APPROACH

CHAPTER 1

DECISION MAKING UNDER UNCERTAINTY

In order to provide background for applications of decision theory, relevant concepts and techniques are described briefly in Chapters 1 and 2. At the outset, attention is given to the nature of decision making and uncertainty. Several decision criteria that have been advanced for choice under uncertainty are discussed. The Bayesian approach is then outlined, and the decision tree technique for structuring decision problems is described.

DECISION MAKING

Decision making may be thought of as a process that includes the following steps:

1. Recognition of a situation that calls for a decision about what action should be taken
2. Identification and development of alternative courses of action
3. Evaluation of the alternatives
4. Choice of one of the alternatives
5. Implementation of the selected course of action

This book concentrates on the first four steps, with steps 3 and 4 receiving the main emphasis because they are the ones for which Bayesian decision theory is most directly applicable.

UNCERTAINTY

Business decisions typically must be made under uncertainty. In other words, doubt exists about what should be done because there is a lack of knowledge of what the outcomes of alternative courses of action would be if they were undertaken.

Uncertainty, as the term is used here, may be contrasted with *risk.* Under risk, the outcome in a given case is not known in advance, but the possible outcomes and their probabilities are known. (In the tossing of a "true" coin, for example, it is "known" from examination and experience that there are two possible outcomes, "heads" and "tails," and that they have equal chances of happening.) Under uncertainty, the full range of possible outcomes may not be apparent, and the probabilities of the possible outcomes are not known in an actuarial sense.

It is uncertainty rather than risk with which the marketing executive usually must contend as he makes decisions relative to products, packages, promotional efforts, prices, channels of distribution, and the like. Working in a continually changing environment, he typically is confronted with unique situations. His experience may give him guidance, but this guidance is limited. He may find it difficult to identify courses of action that deserve serious consideration. Even if the alternatives are seen clearly, their outcomes are likely to be obscure because they represent the net product of a variety of influences. The decision maker may be in a position to shape some of the influential factors (such as product characteristics, price, and the amount of sales effort), but others (such as economic conditions, consumer tastes, and competitors' actions) are beyond his control. He frequently has little information on which to base predictions of how the various factors will affect the outcome of a given program of action at the time it is undertaken.

DECISION CRITERIA FOR UNCERTAINTY

Recognizing the perplexity and pervasiveness of uncertainty, how should the decision maker attempt to deal with it? This question has been answered in various ways by different persons at different times. Several of the resulting decision rules assume that the decision maker is unable because of ignorance to say anything useful about the probabilities of occurrence of the outcomes he regards as possible for alternative courses of action.[1] These rules will be described here briefly before the Bayesian approach and its advantages are discussed.

[1] For more detailed treatments, see William J. Baumol, *Economic Theory and Operations Analysis* (Englewood Cliffs, N.J.: Prentice-Hall, Inc., 1965), pp. 550–568; and Wroe Alderson and Paul E. Green, *Planning and Problem Solving in Marketing* (Homewood, Ill.: Richard D. Irwin, Inc., 1964), pp. 82–101.

One such rule is the *maximin criterion,* which says that the decision maker should determine the worst that could happen under each alternative course of action and then choose the one that would have the highest minimum payoff. If the choice were between the two actions represented in Table 1–1, for which three states of nature (chance events affecting the payoffs) are assumed, A_1 would be elected because its minimum payoff (3) is larger than that for A_2 (0).

Table 1–1

PAYOFF MATRIX

Courses of Action	States of Nature S_1	S_2	S_3
A_1	99	30	3
A_2	90	60	0

At the other extreme is the *maximax criterion,* which dictates choice of the act with the highest maximum payoff. A_1 also would be chosen under this criterion because its largest payoff (99) is greater than that for A_2 (90).

Neither maximin nor maximax takes into account the cost of opportunity lost by making the wrong decision. The *minimax regret* criterion formulated by Savage[2] focuses on opportunity loss. In order to apply it, the decision maker must estimate the possible payoffs for each action. For each state of nature, he then determines the regret that would be suffered if an action other than the one providing the highest payoff were taken.

In the example in Table 1–1, if the first state of nature (S_1) were realized, no regret would be incurred by following A_1, which provides a higher payoff (99) than A_2 does (90). If A_2 were chosen, however, there would be a regret of 9 (99 minus 90). By making similar calculations, the regret matrix in Table 1–2 was developed. Under the minimax regret rule, the maximum regret under each state of nature is identified for each action under consideration. The maximum regret figures in this example are 30 for A_1 and 9 for A_2. Hence, A_2 is chosen because its maximum regret or opportunity loss under any one of the alternative states of nature is less than that for A_1.

The three decision criteria just described ignore most of the information in the payoff matrix as they focus on either the best or the worst that can happen. Another decision rule, the *Laplace criterion,* uses all the information by directing the decision maker to assign equal probabilities to the possible payoffs for each action. A weighted average of the

[2] L. J. Savage, "The Theory of Statistical Decision," *Journal of the American Statistical Association,* no. 46 (1951): 55–67.

Table 1–2

REGRET MATRIX

Courses of Action	States of Nature S_1	S_2	S_3	Maximum Regret
A_1	0	30	0	30
A_2	9	0	3	9

payoffs then is computed for each action by multiplying the payoffs by their respective probabilities and summing the products. The decision rule is to choose the action with the highest weighted average. In the example in Table 1–1, A_2 would be chosen under the Laplace criterion because it has the higher weighted average as follows:

$$A_1 \frac{1}{3} \cdot 99 + \frac{1}{3} \cdot 30 + \frac{1}{3} \cdot 3 = 44$$

$$A_2 \frac{1}{3} \cdot 90 + \frac{1}{3} \cdot 60 + \frac{1}{3} \cdot 0 = 50$$

All four rules discussed above require the decision maker to identify appropriate courses of action and estimate payoffs, yet they assume his complete ignorance concerning their probabilities of occurrence. Actually, if the decision maker can do the former, he probably has information that would lead him to regard some of the possible outcomes as more likely to occur than others. Partial rather than complete ignorance in this regard is characteristic of the business world, in which the decision maker is expected to use his experience and judgment in deciding what to do. The limitation of the complete-ignorance assumption can be lifted by asking the decision maker to assign subjective probabilities—which brings us to the Bayesian approach.

THE BAYESIAN APPROACH

Bayesian decision theory offers an approach for maximizing the executive's chances of achieving the most successful performance of which he is capable, given the limitations of available information and his experience and judgment. It recognizes that in the face of uncertainty, he must gamble. Therefore, it outlines a strategy for making choices that represent the best possible bets. Best bets do not guarantee against loss, of course, but they lead to better performance than poor bets do.

The Bayesian approach suggests that the executive follow these steps:

1. Identify the objectives toward which the decision making should be directed.

2. Identify the alternative courses of action that should be considered.

3. Identify the possible events (environmental conditions) that would influence the payoff of each course of action.

4. Assign a numerical value to the payoff of each course of action, given each possible event.

5. Assign a numerical weight (probability) to the occurrence of each possible event.

6. Using the weights (probabilities), compute the weighted average (expected value) of the payoffs assigned to each course of action.

7. Assess the exposure to both gain and loss associated with each course of action.

8. Choose among the alternative courses of action on the basis of the combination of (*a*) expected value and (*b*) exposure to gain and loss that is most consistent with the decision maker's objectives and attitude toward risk.

This approach should have a familiar ring to many business executives because it formalizes what they have been doing, at least to some extent, intuitively and informally. Use of the more formal procedure should facilitate logical analysis, especially of complex problems of choice.

Brief explanations of some of the terms and steps may be helpful at this point.

OBJECTIVES. The executive may wish to satisfy multiple objectives that vary in kind and specificity. Objectives for a firm, for example, might be profit maximization or maintenance of a given share of the market. The objectives of an individual decision maker probably include favorable evaluation by his superior, something that may go beyond conventional measures of performance. The handling of multiple objectives is discussed later in the book.

COURSES OF ACTION. A course of action usually consists of a sequence of related activities to be taken within a prescribed period of time. A course of action for introducing a new product, for example, would include specifications of the nature, timing, and location of the introductory efforts.

POSSIBLE EVENTS. The word *event* is used to refer to any of a number of environmental factors that may influence the consequences of action but that can not be controlled by the decision maker. (*State of nature* also is used frequently to refer to a condition of the environment.) Two examples of alternative events are: a price cut versus no change in price by a major competitor and a favorable reaction versus an unfavorable reaction to a new product by potential customers. Certain precautions must be taken in listing possible events within a single set. The events must be collectively exhaustive; that is, the list (the set) must be complete in the sense that one of the events on it would occur. Also, the events must be mutually exclusive; that is, they must be defined so that only one of the events on the list can occur.

SUBJECTIVE PROBABILITIES. The distinctive feature of the Bayesian approach is that it calls for the decision maker to quantify his feelings about uncertainty in the form of judgmentally assigned probability distributions for possible outcomes. Once he has done this, uncertainty problems can be analyzed as if they were risk problems for which relative frequency probabilities are available from past performance, perhaps in experimental tests. By requiring subjective probabilities, the Bayesian approach draws on the executive's judgment, conditioned as it is by past experience, for whatever value it may have for the task at hand. In so doing, it departs from traditional statistics, which was limited to the relative-frequency view of probability. Bayesian decision theory provides for the use of both managerial judgment and statistical evidence.

EXPECTED VALUE. How would you value a course of action that has a .6 chance of resulting in a gain of \$400 and a .4 chance of resulting in a loss of \$200? For the moment, assume that you are able and willing to take any chance as long as it is worth it monetarily. In order to determine the worth in this case, you would weight each of the possible outcomes for its probability of occurrence and sum the products.

$$\begin{aligned} .6 \times +\$400 &= +\$240 \\ .4 \times -\$200 &= \underline{\quad -80} \\ & +\$160 \end{aligned}$$

The answer of \$160 is the *expected value* of the course of action. Inasmuch as it is expressed in terms of dollars, it is the *expected monetary value*. The expected-value concept, however, is applicable regardless of the kind of value used.

Note that expected value is not the most likely outcome. The most likely outcome in this example is +\$400. Normally you would be unwilling to pay that much for the opportunity to take the course of action because the latter carries with it a .4 chance of a \$200 loss. For purposes of evaluating the alternative, then, we employed expected value, which is an average in which all possible outcomes are weighted by their probabilities. In the above example, you would be warranted in paying up to (but not more than) \$160 for a chance to take the course of action on the basis of expected monetary value.

EXPOSURE TO GAIN AND LOSS. For many business situations, expected monetary value is an adequate criterion for choice. In others, however, it is not. For example, assume that you must adopt one of the following two alternatives:

	Possible Outcomes	Probability	Expected Monetary Value
Alternative *A*	+$400	.6	+$240
	−100	.4	−40
			+$200
Alternative *B*	+$1,000	.5	+$500
	−600	.5	−300
			+$200

Which of the alternatives would you choose? They are equal in expected monetary value. Nevertheless, you probably find that you are not indifferent. You note that the alternatives involve different exposures to both gain and loss. Your choice will depend on the attitude toward risk that you have in your current circumstances. If you are anxious to avoid loss, you will select alternative *A*. On the other hand, if an opportunity to realize a large gain has unusual attraction, you might select alternative *B*, even though you are more likely to incur a loss and the amount of loss would be greater than under alternative *A*. If your desire for large gain were great enough, you would choose alternative *B* even if its expected monetary value were lower than that of the more conservative alternative.

In choosing among courses of action, the decision maker should be aware of the range of the possible payoffs and their odds, as well as the expected values. When the differences in exposure to gain and loss are substantial enough to affect the choice, they should be taken into account. One means of doing this, for which there are formal procedures, consists of converting monetary values into *utility* values that allow for the decision maker's attitude toward risk.[3] The alternative with the highest expected utility can then be chosen. Another approach that has certain practical advantages in working with business executives was illustrated

[3] For treatments of utility theory, see Howard Raiffa, *Decision Analysis: Introductory Lectures on Choices under Uncertainty* (Reading, Mass.: Addison-Wesley Publishing Co., Inc., 1968), chapter 4; Robert Schlaifer, *Probability and Statistics for Business Decisions* (New York: McGraw-Hill Book Company, 1959), chapter 2; Ralph O. Swalm, "Utility Theory—Insights into Risk Taking," *Harvard Business Review* 44, no. 6 (November–December 1966): 123–136; John S. Hammond, III, "Better Decisions with Preference Theory," *Harvard Business Review* 45, no. 6 (November–December 1967): 123–141.

in the above example. This approach consists of reviewing both the probability distributions of the possible outcomes of the courses of action under consideration and their expected values. A more elaborate application of this approach appears in Chapter 10.

A SIMPLE ILLUSTRATION: BAYESIAN PRIOR ANALYSIS

We now shall elaborate on the brief description of the Bayesian approach by putting it to work in an artificially simple example (before approaching the complications of real situations presented in later chapters). Assume that the marketing manager for a soft drink is considering whether to undertake a special promotion for his product in a large market area during the month of June and that he must make a decision now.

Table 1–3

EXPECTED PAYOFFS: SOFT-DRINK SPECIAL PROMOTION

Possible Consumer Reactions	Alternative Courses of Action		Probabilities of Consumer Reactions
	A_1	A_2	
Very favorable	\$300,000	\$0	.4
Favorable	100,000	0	.3
Unfavorable	−200,000	0	.3
Expected payoffs	\$ 90,000	\$0	1.0

Following the steps of the Bayesian approach, the manager specifies profit maximization as his sole objective and decision criterion. Under the constraints of this example, he has only two alternatives: (1) he can approve the plans for the special promotion, or (2) he can reject them. In considering the possible payoffs, the manager has noted that the special promotion would cost \$200,000. No objective evidence was available on its probable sales effectiveness. He believes, however, that the range of possible consumer reactions could be meaningfully represented by three categories: "very favorable," "favorable," and "unfavorable." Submitting to the rigors of the Bayesian procedure, he summarizes his feelings about the possible economic consequences of each of the categories in the form of a payoff table (see Table 1–3).[4] All payoffs are net after deducting the cost of the special promotion.

The marketing manager then proceeds to calculate the expected mone-

[4] The number of payoffs chosen to represent the range of possibilities could have been larger than three. In fact, it is technically possible to work with the entire distribution of possible outcomes. For the purposes of this book, however, we shall work with the smallest number of values believed necessary to adequately represent the total distribution.

tary values of payoffs. The expected value of the second alternative is \$0. The expected value of the first alternative is (\$300,000 × .4) + (\$100,000 × .3) + (−\$200,000 × .3) = \$90,000.

The same result can be reached by comparing the expected opportunity losses of the alternatives. The *expected opportunity loss* is the difference between the cost or profit that would be realized under a particular decision and the cost or profit that would have been realized had the decision been the best one.

The expected opportunity loss of an alternative is computed by multiplying the possible opportunity losses by their prior probabilities. In Table 1–4, under alternative A_1, "very favorable" or "favorable" consumer reactions to the special promotion would bring no loss, but an unfavorable reaction would result in a loss of \$200,000. The probability of an unfavorable reaction was judged to be .3; therefore the expected loss for A_1 is (\$200,000 × .3) = \$60,000. If A_2 were chosen, the company would lose the opportunity it would have had under A_1 for gains of \$300,000 if consumer reaction were "very favorable" and \$100,000 if it were "favorable." If the reaction were "unfavorable," no loss would result. The expected opportunity loss for A_2 is (\$300,000 × .4) + (\$100,-000 × .3) + (\$0 × .3) = \$150,000. The difference between the expected opportunity losses for the two alternatives is (\$150,000 − \$60,000) = \$90,000, the expected monetary value of A_1 that we found earlier by calculating the difference between the expected payoffs of A_1 and A_2. The difference between the expected payoffs of two alternatives will always be the same as the difference between their expected opportunity losses, a necessary consequence of the procedures used in their derivation.

Table 1–4

EXPECTED OPPORTUNITY LOSSES: SOFT-DRINK SPECIAL PROMOTION

Possible Consumer Reactions	Alternative Courses of Action		Probabilities of Consumer Reactions
	A_1	A_2	
Very favorable	\$ 0	\$300,000	.4
Favorable	0	100,000	.3
Unfavorable	200,000	0	.3
Expected opportunity loss	\$ 60,000	\$150,000	1.0

Given the constraint that the decision must be made now, the manager would choose A_1 (run the special promotion) to take advantage of its expected payoff of \$90,000 or, to put it another way, to avoid the expected opportunity loss of \$90,000 associated with A_2.

This is an illustration of *prior analysis,* so-called because it was made on the basis of the manager's present judgments, prior to obtaining any additional information. In this example, it was assumed that circumstances precluded undertaking alternatives designed to produce evidence of the probable sales effectiveness of the promotion. In Chapter 2, this assumption will be abandoned in order to illustrate how prior judgments can be revised to reflect additional information obtained after the prior judgments have been made.

DECISION TREES

The decision tree concept can facilitate the structuring and discussion of decision problems. It is introduced now because it is used in the illustrations of the Bayesian approach throughout this book.

The decision tree is a means of displaying the anatomy of a decision.[5] It consists of a series of nodes and branches. Each alternative course of action under consideration is represented by a main branch which, in turn, may have subsidiary branches for related chance events that appear in chronological sequence. In other words, the tree diagrams the paths that lead to the possible consequences. In addition to the structure itself, the tree may show the payoffs for each path and the probabilities for the various chance events.

A decision tree for the soft-drink special-promotion problem that was presented earlier would have two main branches. For the A_1 branch (run the special promotion), the marketing manager had designated three chance alternatives in terms of consumer reactions ("very favorable," "favorable," and "unfavorable") for which he estimated economic consequences of +$300,000, +$100,000, and −$200,000, respectively. For A_2 (reject the special promotion), he assumed there would be no change from normal in consumer buying of the company's soft drink that could be attributed to the decision not to run the special promotion. The problem is structured in decision tree form in Figure 1–1.

It is common (though not universal) practice to represent decision points by squares and chance alternatives by small circles. This is a means of distinguishing between action choices over which the decision maker has control and events that are beyond his control. In Figure 1–1, probabilities for the chance events of consumer reaction are given in parentheses and the expected monetary values (EMV) of the courses of action appear at the chance alternative nodes. Inasmuch as the first alternative course of action has the larger expected monetary value, it is chosen. The second alternative is rejected, as indicated by the double slash marks drawn through the A_2 branch.

[5] For more detailed presentations, see John F. Magee, "Decision Trees for Decision Making," *Harvard Business Review* 42, no. 4 (July–August 1964): 126–138; idem, "How to Use Decision Trees in Capital Investment," *Harvard Business Review* 42, no. 5 (September–October 1964): 79–96.

The decision tree can be of great value in helping management achieve an overall picture of the decision situation, one that recognizes the possible action choices, the related risks, and the possible outcomes. It also can serve to clarify the decision criteria and the information needed for decision-making purposes.

The utility of the decision tree technique becomes more apparent as the decision problem becomes more complicated. For purposes of exposition, the special-promotion example was limited to one chance event, that is, consumer reaction to the special promotion. The consumer re-

FIGURE 1-1

DECISION TREE: SOFT-DRINK SPECIAL-PROMOTION PROBLEM

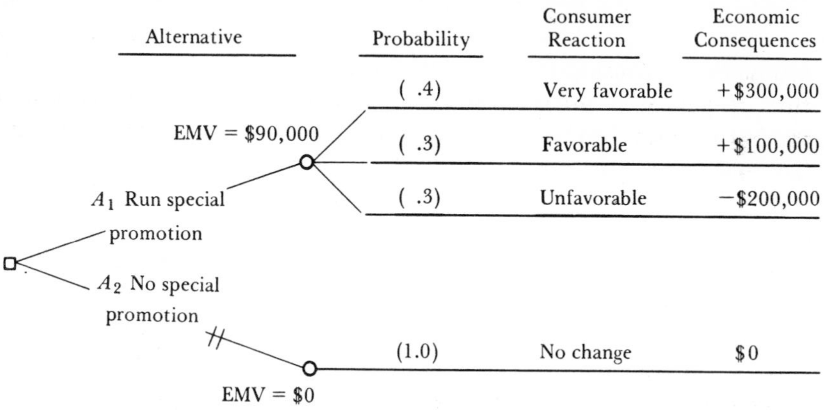

Alternative	Probability	Consumer Reaction	Economic Consequences
A_1 Run special promotion (EMV = $90,000)	(.4)	Very favorable	+$300,000
	(.3)	Favorable	+$100,000
	(.3)	Unfavorable	−$200,000
A_2 No special promotion (EMV = $0)	(1.0)	No change	$0

action, however, probably would vary with the temperature and with the amount of competitive promotional activity. It might be helpful to the marketing manager if these two additional factors were represented in the decision tree so that they would be more visible. Their inclusion could encourage more specific thinking. In this case, for example, the manager might be reminded that rejection of the proposal could lead to economic loss if a major competitor stepped up his promotional efforts.

The decision tree in Figure 1–2 incorporates these factors. Inasmuch as our objective at this point is only that of illustrating a more elaborate framework, the values needed for the analysis are not filled in. It should be apparent, however, that the probabilities and possible outcomes will vary for the different circumstances represented by branches of the tree.

The main criteria for a decision tree are that it be suitable to the problem and helpful to the decision maker. There is no one best way to lay out a tree. It should be limited, however, to decisions and events that have consequences the decision maker wishes to compare. It should include sufficient detail to sharpen his thinking but not so much that he feels inundated and fails to focus on the key issues.

The time period embraced by the tree will, of course, vary with the problem. It could be a few days or several years. In a given case, the period chosen for the analysis should extend to a point beyond which it can be assumed that further changes will not occur in the differences between the outcomes of the alternative courses of action being evaluated. If the time period is sufficiently long and the costs and payoffs of the alternatives vary substantially in amount and timing, the values should be discounted for the cost of capital so that the alternatives will be on a

FIGURE 1-2

A MORE ELABORATE DECISION TREE STRUCTURE: SOFT-DRINK SPECIAL-PROMOTION PROBLEM

Action Alternatives	Chance Alternatives		Consumer Response[1] Category	Probability	Economic Consequences[2]
A_1 Run special promotion	Weather favorable ( )	Normal competitive promotion ( )	Very favorable	( )	$
			Favorable	( )	$
			Unfavorable	( )	$
		Increased competitive promotion ( )	Very favorable	( )	$
			Favorable	( )	$
			Unfavorable	( )	$
	Weather unfavorable ( )	Normal competitive promotion ( )	Very favorable	( )	$
			Favorable	( )	$
			Unfavorable	( )	$
		Increased competitive promotion ( )	Very favorable	( )	$
			Favorable	( )	$
			Unfavorable	( )	$
A_2 No special promotion	Weather favorable ( )	Normal competitive promotion ( )	No change	( )	$
		Increased competitive promotion ( )	Very favorable	( )	$
			Favorable	( )	$
			Unfavorable	( )	$
	Weather unfavorable ( )	Normal competitive promotion ( )	No change	( )	$
		Increased competitive promotion ( )	Very favorable	( )	$
			Favorable	( )	$
			Unfavorable	( )	$

1–Response is to the company's special promotion under A_1 but to competitive promotion under A_2.
2–Economic consequences for the company faced with the decision. Probabilities to appear in the parentheses.

comparable basis. The alternatives then would be compared by their discounted expected values.

Although the decision tree concept itself is simple, its implementation requires the skill and judgment that come from experience. In a good many business decisions, structuring is the most difficult and most important part of the process. Once a clear structure is developed, the best decision may be apparent even without detailed computations. Structuring will be given substantial attention later in the book, especially in the context of the freeze-dried coffee decision problem.

STUDY QUESTIONS

1. For what steps in the decision process is the Bayesian approach directly applicable?

2. Distinguish between *uncertainty* and *risk* as the terms were used in this chapter.

3. Describe the Bayesian approach to choice under uncertainty.

4. Explain the terms *expected value* and *exposure to gain and loss.* Would you use one of these measures or both in choosing among alternatives? Why?

5. Describe the main features of a decision tree. What considerations should govern its design?

CHAPTER 2

THE COST AND VALUE OF INFORMATION

With the development of Bayesian statistics, business for the first time was provided with a theoretical framework for combining managerial judgment and new data in decision making. The procedure for making choices on the basis of judgment alone was described in Chapter 1. This chapter is devoted to the valuing of information in order to arrive at answers to the following questions:

1. Should an attempt be made to obtain additional information before making a final decision on a course of action? If so, what expenditure of funds would be warranted?

2. How should management evaluate alternative research proposals?

3. How should new information be combined with managerial judgment so that both may have appropriate influence on the decision?

SEEK MORE INFORMATION?

In many situations, management has the option of ordering inquiries before taking final action. The resulting informational inputs for the decision making would have value to the extent that they reduced uncertainty and increased the likelihood that the best course of action would be chosen. The inputs, however, come at a cost. The collection and analysis of data involve expense. In addition, opportunity for gain may be lost by delaying final action until the research can be completed. Of course, management would prefer to order research only if the findings could reduce uncertainty enough to warrant the costs involved. Bayesian decision theory provides a means for applying this criterion. In describing the procedure, we shall return to the soft-drink special-promotion decision problem introduced in Chapter 1 and assume that the marketing manager can seek additional information before taking final action. He must decide whether he should do so.

TOP LIMIT FOR RESEARCH EXPENDITURE. Before evaluating proposed inquiries, it is possible to determine a research cost that should not be exceeded. It is the value of *perfect information,* that is, information which would permit choice of the best action with certainty. This value can be arrived at by two routes that represent different views of the situation.

One approach is that of determining the expected opportunity loss of the alternative identified as the best one by prior analysis. In the example, expected opportunity loss is the difference between the cost that would be incurred by following A_1 and the lesser cost that would result from following A_2 if consumer reaction to the special promotion were unfavorable.

In the absence of additional information, the marketing manager would choose A_1 because its expected value was \$90,000 compared with an expected value of \$0 for A_2 (see Table 1–3). Assume for the moment that he could buy research which would tell him for certain whether consumers would react positively to the proposed special promotion. How much should he be willing to pay for it?

If the manager were to learn for certain that consumer reaction would be favorable, he would choose A_1, but he would have done that anyway. In this event, the purchase of perfect information would increase his cost but not his payoff. If he were informed that consumer reaction would be unfavorable, however, he would choose A_2 to avoid a \$200,000 loss that would be incurred under A_1. That loss, of course, must be evaluated in advance of knowing whether the perfect information would reveal a favorable or an unfavorable consumer reaction. On the basis of the marketing manager's prior judgment, the probability of his learning of an unfavorable consumer reaction is .3. The expected loss under A_1, therefore, is (\$200,000 × .3) = \$60,000 (see Table 1–4). A_2 would involve no loss under the assumption of an unfavorable consumer reaction to the special promotion. The expected opportunity loss associated with A_1, therefore, is (\$60,000 – \$0) = \$60,000. The latter figure represents the *cost of uncertainty* of the decision situation.

The same conclusion can be reached by comparing the expected value under uncertainty of the alternative judged best by prior analysis with the expected value of that alternative under certainty. If the marketing manager were to act without the benefit of more information, he would choose A_1, which has an expected value of $90,000 (see Table 1–3). Upon receiving perfect information, he would choose A_1 if he were told that consumer reaction would be positive and A_2 if he were told that it would be unfavorable. He could look forward, therefore, to payoffs of $300,000 with probability .4, $100,000 with probability .3, and $0 with probability .3, for an expected monetary value of $150,000 (see Table 2–1). The incremental expected monetary value of getting perfect information is ($150,000 – $90,000) = $60,000. Guided by the expected-value criterion, the manager should be unwilling to spend more than $60,000 for research.

Table 2–1

EXPECTED PAYOFFS UNDER CERTAINTY
SOFT-DRINK SPECIAL-PROMOTION PROBLEM

Possible Consumer Reactions	Alternative Courses of Action		Probabilities of Consumer Reactions
	A_1	A_2	
Very favorable	$300,000	$0	.4
Favorable	100,000	0	.3
Unfavorable	0	0	.3
Expected payoff	$150,000	$0	1.0

The assumption of the availability of perfect information, of course, was unrealistic. Research findings are subject to errors because of the limitations of methods of gathering and analyzing data. Even with error-free findings, current knowledge does not permit perfect prediction of outcomes of future actions. The maximum the marketing manager should spend for research, therefore, is less than $60,000. The exact amount depends on the value of the imperfect information that would be produced by research.

VALUING INFORMATION

The decision maker seeks additional information when confronted with alternative courses of action in order to revise his prior probabilities so that he will be able to arrive at better expected values to guide his choice. Revisions of prior probabilities were implicit in our computations of the value of perfect information. Under certainty, the probability of either a very favorable or a favorable consumer reaction to the special promotion was judged by the marketing manager to be .7. Under an assumption of certainty, it became 1.0 by definition. The possible reactions of "very favorable" and "favorable" remained in a 4 to 3 relationship to one another in their likelihood of occurrence. Similarly, when an assumption of certainty was made, the probability of an unfavorable consumer reaction changed from .3 to 1.0.

The appropriate revision of prior probabilities to reflect imperfect information, however, is not so readily apparent. Assume that you had judged the probability to be .6 that the Democratic candidate for governor would be elected in your state. Three weeks before the election, you learn that a statewide public opinion poll showed that registered voters preferred the Democratic candidate over his opponent by a 4 to 3 margin. You are confident that the poll was well conducted, but you know that polls are not perfect and that the election still is three weeks away. Should you revise your prior probability for the election of the Democratic candidate? If so, how much of a revision should you make?

Bayesian decision theory provides for incorporating additional information into the analysis by revising prior judgments in accordance with probability theory.

PROBABILITY THEORY

Before illustrating this procedure for revising prior judgments, we shall discuss some of the main features of probability theory. It should be clear from our earlier discussion that probability is an expression of one's confidence in the occurrence of something. Although the feeling may be based on recorded past performance, it need not be. In order to use probability theory in indirect assessments of probabilities in the Bayesian approach, these feelings must be expressed in terms of weights or probabilities that are assigned in accordance with the following axioms:

1. A probability is a real number between 0 and 1 assigned to an event.

2. The sum of the probabilities assigned to a set of mutually exclusive and collectively exhaustive events must be 1.

3. The probability of an event that consists of a group of mutually exclusive subevents is the sum of their probabilities.

CONDITIONAL, JOINT, AND MARGINAL PROBABILITIES. The probability assigned to an event when the occurrence of another event is known or assumed is the *conditional* probability of the first event, given the second. For example, you might assign a probability of .7 to the election of the Democratic candidate for governor, conditional upon the election of the Democratic candidate for United States senator.

The probability that two or more events will occur is the *joint* probability of the events. For example, you might assign a joint probability of .4 to the election of both the Democratic candidates for governor and United States senator.

The probability assigned to a particular event independent of any other event is referred to variously as the *unconditional, ordinary, simple,* or *marginal* probability. For example, all things considered, you might assign a probability of .6 to the election of the Democratic candidate for governor. The unconditional probability of an event can be obtained by adding the probabilites of all joint events of which it is a part. When it is computed in this way (by adding down a column or across a row of a joint probability table) it appears as a sum in the margin and is referred to as the marginal probability of the event.

In applying Bayesian decision theory, we shall be concerned with the relationships between conditional, joint, and unconditional (marginal) probabilities. In this connection, the addition and multiplication rules of mathematical probability are relevant.

THE ADDITION RULE. Assume that you wish to calculate the probability that either one or the other of two mutually exclusive events (A or B) will occur. The addition rule tells you to add their unconditional probabilities to get the answer.[1]

$$P(A \cup B) = P(A) + P(B)$$

If A and B are two events that are not necessarily mutually exclusive, the probability of either one or the other occurring can be found by adding their unconditional probabilities and subtracting the probability of their joint occurrence.

$$P(A \cup B) = P(A) + P(B) - P(A \cap B)$$

THE MULTIPLICATION RULE. The multiplication rule tells you how to calculate joint probabilities. In the case of two independent events (A and B), the probability of both of them occurring is the product of their unconditional probabilities.

$$P(A \cap B) = P(A) \cdot P(B)$$

The concept of conditional probability is needed for the computation of the joint probability of two events that are not independent. The joint probability of two events (not necessarily independent) is calculated by multiplying the conditional probability of one event, given the occurrence of the second event, by the unconditional probability of the second event.

$$P(A \cap B) = P(A|B) \cdot P(B)$$

In the interest of simplicity, only two events were used in the above examples. The formulas, however, can be extended to cover more than two events.

[1] In formulas that follow, the symbol $\cup$ stands for "or" and the symbol $\cap$ stands for "and." The conditional probability of an event (A), given the occurrence of another event (B), is written as $P(A|B)$.

CALCULATING CONDITIONAL PROBABILITY. If you are to revise a prior probability to reflect new information, you need a way of calculating the probability of the event in question, given the new information. The mathematical definition of conditional probability is expressed by the following formula:

$$P(A|B) = \frac{P(A \cap B)}{P(B)}$$

The definition tells us that the probability of an event (A), given the occurrence of a second event (B), can be obtained by dividing the probability of their joint occurrence by the unconditional probability of the second event (B). The definition itself, however, does not tell us how to compute the values needed for the numerator and the denominator. This information is provided by Bayes' theorem.[2] It supplements the above definition by specifying how the needed calculations can be made, employing the addition and multiplication rules. The detailed expression of the theorem by formula will vary with the available information and requirements of the problem at hand. One expression of the theorem follows:

$$P(A|B) = \frac{P(A) \cdot P(B|A)}{P(A) \cdot P(B|A) + P(A') \cdot P(B|A')}$$

To interpret the formula, let A stand for the occurrence of a particular event, A' for the nonoccurrence of the same event, B stand for a particular observation (new information) relative to the occurrence of A. The numerator specifies the calculation of the joint probability that A would occur and that observation B would be made. The denominator specifies the calculation of the probability of observation B. That probability is the sum of the probability that observation B would be made if A would occur and the probability that observation B would be made even if A would not occur.

In Bayesian analysis, Bayes' theorem is frequently, but not always, used in revising prior probabilities in the light of new information. An alternative procedure that is based on the same concepts but does not employ the theorem itself will be described in Chapter 5.

AN EXAMPLE OF USE OF BAYES' THEOREM: POSTERIOR ANALYSIS

Let us return to a question raised a few pages earlier: How should you revise your prior probability .6 for the election of the Democratic

[2] Reverend Thomas Bayes, "An Essay Toward Solving a Problem in the Doctrine of Chances," *Philosophical Transactions,* 1764. (Bayes was a clergyman and a teacher of mathematics who became interested in probability theory. The label "Bayesian" recognizes his work on the revision of subjective probabilities based on observations. Bayesian statistics as we know it today, however, was developed primarily in the past dozen years by the contributions of a number of scholars.)

candidate for governor upon learning three weeks before the election that a public opinion poll found that voters in the state preferred him by a 4 to 3 margin over his opponent? The question, of course, calls for the conditional probability of the candidate's election, given the poll result. You know from Bayes' theorem that you can get the answer by using the following formula:

$$P(\text{C's election}|\text{poll outcome}) = \frac{P(\text{C's election}) \cdot P(\text{poll outcome}|\text{C's election})}{P(\text{poll outcome})}$$

You already have your prior probability .6 of the candidate's election. You now need to assign the probability that the poll would show a 4 to 3 preference for the candidate for the time it was taken if the candidate were going to win on election day three weeks hence. Making this judgment, of course, is not easy. You are aware that poll results are subject to some sampling and systematic error; therefore, the reported 4 to 3 preference may not be an accurate representation of voter preference in the state at that time. Although the poll was limited to registered voters, those polled may not be representative of the voters who actually get to the polls. In addition, preferences may have changed by election day. We shall assume, however, that you have considered these and other factors and have concluded that the probability was .7 for obtaining the 4 to 3 preference in the poll if the candidate were going to win (see Table 2–2). (For purposes of illustration, we shall limit the example to two categories of poll results: (1) at least a 4 to 3 preference for the candidate and (2) less than a 4 to 3 preference for the candidate.

Table 2–2

CONDITIONAL PROBABILITIES OF POLL OUTCOMES: GUBERNATORIAL ELECTION EXAMPLE

	Poll Outcomes	
Election Outcomes	*≥4–3 Preference for Candidate*	*<4–3 Preference for Candidate*
Candidate wins	.7	.3
Candidate loses	.3	.7

You can now calculate the joint probability of the candidate's election and the reported poll result. Simply multiply your prior probability of his election by the conditional probability of obtaining the reported poll outcome if the candidate were going to win (see Table 2–3).

The marginal probability of the reported poll result is the sum of the joint probabilities in which it is involved; therefore, it is obtained by adding the joint probabilities in the first column of the table. You now

Table 2–3

JOINT AND MARGINAL PROBABILITIES OF ELECTION RESULTS AND POLL OUTCOMES: GUBERNATORIAL ELECTION EXAMPLE

	Poll Outcomes		
Election Outcomes	≥ 4–3 Preference for Candidate	< 4–3 Preference for Candidate	Marginal Probabilities
Candidate wins	.6 × .7 = .42	.6 × .3 = .18	.60
Candidate loses	.4 × .3 = .12	.4 × .7 = .28	.40
Marginal probabilities	.54	.46	1.00

have the values required for the computation of the conditional probability of the candidate's election, given the reported outcome of the poll

$$P(\text{C's election}|\text{poll outcome}) = \frac{P(\text{C's election} \cap \text{poll outcome})}{P(\text{poll outcome})} = \frac{.42}{.54} = .78$$

Having observed the poll result, you should feel more certain than you were that the candidate is going to win and revise your prior probability of his election upward to .78. This is a limited example of *posterior analysis,* in which prior judgments are revised by incorporating new information after it has been obtained.

EVALUATING A RESEARCH PROPOSAL: PREPOSTERIOR ANALYSIS

At this point, we shall return our attention to the marketing manager of the soft drink company who must decide whether to run a proposed special promotion. By prior analysis, he found that he should do so if he had to make the decision immediately. In considering whether he should seek more information before acting, it was determined that he should not spend more than $60,000 for research, that figure representing the value of perfect information. To go further in answering the question, it is necessary to evaluate specific research plans.

For purposes of illustration, we shall examine only one simplified proposal. Alternative plans might be compared, but each would be subjected to the same treatment. The analytical approach was suggested by the posterior analysis in the gubernatorial election example. In that case, however, new information already had become available. The task was to evaluate it and revise prior judgments accordingly. The marketing manager must decide whether to order an inquiry before its results can be known. The decision requires that he judge what the findings might be and then determine whether they would be worth the cost. The analysis involves the following steps:

1. List the possible research outcomes and calculate their marginal probabilities.

2. Assume, in turn, that each of the research outcomes has been obtained. For each research outcome:
   a. Revise the prior probabilities.
   b. Calculate the expected payoff of each course of action under consideration and select the act with the highest expected payoff.
   c. Multiply the expected payoff of the best course of action by the marginal probability of the research outcome.

3. Sum the products of step 2c (taken for each research outcome) to get the expected payoff of the strategy that includes ordering research before taking final action.

4. Subtract the cost of the research to get the expected net payoff of the strategy.

5. Compare the expected net payoff of the strategy that includes research with the expected payoff of the strategy of choosing among the alternatives without research.

6. Choose the strategy that maximizes expected net payoff (assuming the use of the decision criterion of maximum expected value).

The evaluation of a strategy that includes collecting information before taking final action is referred to as *preposterior analysis.*

THE RESEARCH PROPOSAL. At the marketing manager's request, a research firm proposed a market test of the special promotion. It would run for ten days and be completed early enough so that it would not delay the launching of the special promotion. It would cost $3,500.

THE ANALYSIS. The manager began his consideration of whether he should buy the test by structuring the problem in the form of the decision tree shown in Figure 2–1. He would adopt either the strategy of making a final decision without a market test or the strategy that included the test, depending on which had the higher expected payoff.

The manager identified three categories of possible market-test outcomes represented as follows: a 15 percent increase in sales (T_1), a 5 percent increase in sales (T_2), and no sales increase (T_3). He reasoned that if the special promotion, run in June as originally proposed, were to receive a very favorable consumer reaction, the probability of observing T_1 in the test was .6, the probability of observing T_2 was .3, and

FIGURE 2-1

DECISION TREE FOR SOFT-DRINK SPECIAL-PROMOTION PROBLEM: NO MARKET TEST VERSUS MARKET TEST

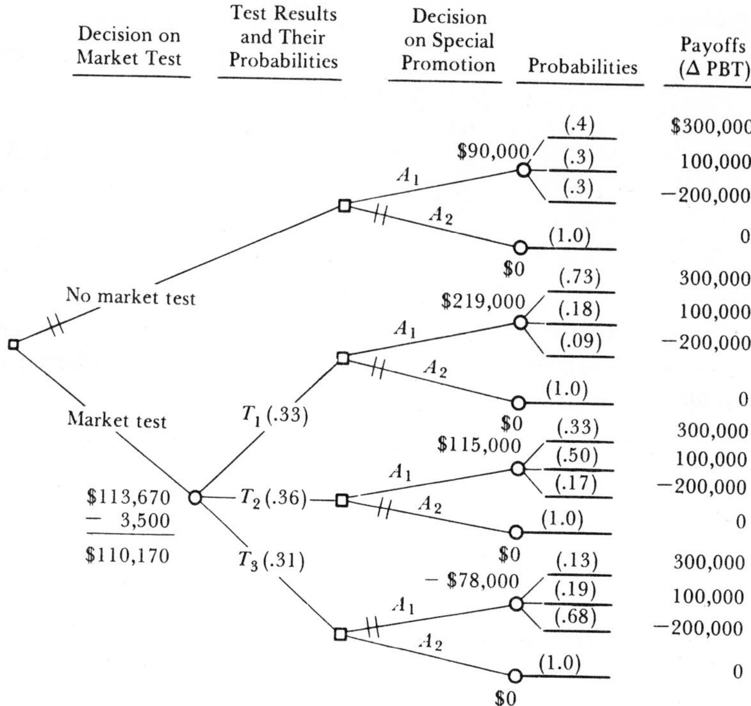

the probability of observing T_3 was .1. Similarly, he assigned conditional probabilities to the possible test outcomes for the other possible states of consumer reaction (see Table 2–4). He then multiplied the conditional probabilities by his prior probabilities to arrive at their joint probabilities (see Table 2–5). For example, his prior probability of .4 for S_1 times the conditional probability of .6 for T_1 given S_1 gave a joint probability for S_1 and T_1 of .24, and so forth.

Table 2–4

CONDITIONAL PROBABILITIES OF MARKET-TEST RESULTS

States of Consumer Reactions to Special Promotion If Run as Proposed	Test Results T_1 (+15 percent)	T_2 (+5 percent)	T_3 (±0)
S_1 (very favorable)	.6	.3	.1
S_2 (favorable)	.2	.6	.2
S_3 (unfavorable)	.1	.2	.7

Table 2–5

JOINT PROBABILITIES OF STATES AND MARKET-TEST RESULTS

States of Consumer Reactions to Special Promotion If Run as Proposed	Test Results T_1 (+15%)	T_2 (+5%)	T_3 (±0%)	Marginal Probabilities
S_1 (very favorable)	.24	.12	.04	.4
S_2 (favorable)	.06	.18	.06	.3
S_3 (unfavorable)	.03	.06	.21	.3
Marginal probabilities	.33	.36	.31	1.00

He next revised his prior probabilities for the possible outcomes of running the special promotion, using the following formula:

$$P(S_i|T_j) = \frac{P(S_i \cap T_j)}{P(T_j)} \qquad (i, j = 1, 2, \ldots)$$

For example, here are three of his calculations.

$$P(S_1|T_1) = \frac{P(S_1 \cap T_1)}{P(T_1)} = \frac{.24}{.33} = .73$$

$$P(S_2|T_1) = \frac{P(S_2 \cap T_1)}{P(T_1)} = \frac{.06}{.33} = .18$$

$$P(S_3|T_1) = \frac{P(S_3 \cap T_1)}{P(T_1)} = \frac{.03}{.33} = .09$$

He made similar calculations under the assumptions that each of the other two possible outcomes were observed in the market test and entered the revised probabilities in the decision tree (see Figure 2–1).

By multiplying the possible outcomes of A_1 by the revised probabilities, he found that the expected payoff of running the promotion would be \$219,000 if T_1 were observed in the test, \$115,000 if T_2 were observed, and –\$78,000 if T_3 were observed. If T_1 or T_2 were observed, he would choose A_1 (run the special promotion); but if T_3 were observed, he would choose A_2 (reject the special promotion). The marginal probabilities of the possible test outcomes were obtained by summing their joint probabilities (see Table 2–5).

To arrive at the expected net payoff of the market-test strategy, he multiplied the expected payoff of the action he would take under each of the possible test outcomes by the probability of observing the test outcome and summed the products: (\$219,000 × .33) + (\$115,000 × .36) + (\$0 × .31) = \$113,670. He then deducted the cost of the research (\$3,500) to obtain the expected net payoff of \$110,170. The latter value exceeded by \$20,170 the expected payoff of the alternative strategy of taking final action without market testing, so he ordered the test. In this case, he would have been warranted in spending up to $20,170, that amount being the expected value of the information which would be produced by the test.

STUDY QUESTIONS

1. What is the value of *perfect information?* Of what practical value is the concept?

2. Distinguish between marginal, conditional, and joint probabilities.

3. How does the Bayesian approach provide for combining executive judgment and new information in decision making?

4. In terms of Bayesian analysis, what conditions must be met to justify ordering research before making a terminal decision?

5. What is the Bayesian approach to choosing among alternative research proposals?

PART II

APPLICATIONS OF THE BAYESIAN APPROACH

In order to gain an overview of the Bayesian approach, we focused on concepts and techniques in Chapters 1 and 2, limiting illustrations to simplified examples. We now shift our attention to the operational aspects of the application of the Bayesian approach in actual business decision situations. In addition to questions of analytical procedure, we shall concentrate on the exercise of managerial judgment for the structuring of complex problems and the determination of values requisite to successful implementation.

Part II is divided into three sections; the first two deal with different decision problems. A major packaging change is considered in Section One. After a description of the decision situation, alternative courses of action that represent terminal decisions based on information already available are analyzed (prior analysis). Attention is then given to evaluating actions that include research to obtain additional information for use in arriving at a final decision (preposterior analysis). The analyses are made first without and then with the aid of the computer. Section Two focuses on a considerably more complex problem that calls for decisions relative to the technical development and brand positioning of a product innovation. Consideration is given to the identification and screening of alternatives as well as to the planning and conduct of a prior analysis by computer. Section Three presents concluding comments on applying the Bayesian approach.

Section One

A Decision Problem for Prior and Preposterior Analysis

CHAPTER 3

CONSIDERING A MAJOR PACKAGING CHANGE: THE DECISION CONTEXT

The product manager for Maxwell House ground coffee was considering what action should be taken now that he had been notified by the American Can Company that developmental work on a container design which employed the tear-strip opening principle was about to be completed and that a supply of the new cans soon could be made available to Maxwell House.

Officials of American Can Company expected to be able to deliver 150,000 of the "quick-strip" cans in the one-pound size in December, 1963, for test marketing. They believed that they could build a production line which could start turning out a million cans a week by the middle of March, 1964, and then steadily increase capacity so that four months later it would be sufficient to supply the new can for the total output of Maxwell House ground coffee.

Packaging and labeling were matters of concern to the product manager because of his responsibility for the domestic marketing of his brand. Aided by a group of assistants, he established annual sales objectives, worked with the advertising agency to develop and implement advertising and promotional programs, evaluated possible changes in the product, determined what price changes should be recommended, and initiated sales tests and other consumer research.

The expected availability of the quick-strip can prompted consideration of changing containers for the second time in a year. Prior to 1963, the packaging of ground coffee in the United States had undergone no major changes since the late 1920s, when a can developed for shortening products was adapted for coffee. At that time, the large can suppliers tooled to produce a key-opening can 5⅛ inches in diameter and 3⅝ inches in height. The can had been used since then by most brands except during World War II, when glass jars were substituted temporarily.

THE KEYLESS CAN

New activity in packaging became apparent in October, 1962, when Folger's began test marketing in Stockton, California, of a keyless container made by the American Can Company. The no-key can was taller, narrower, and less expensive than the traditional key-opening tin. It was made to be opened with a regular can opener and employed a polyethylene lid for reclosure.

In November, 1963, Maxwell House initiated a survey of 125 Stockton users of Folger's coffee in the new container and audits of ground-coffee sales in 21 Stockton retail stores for three weeks. The audits provided no evidence of change in Folger's market share brought about by the keyless can, but the survey showed that 86 percent of the women who had tried it preferred it to the key can. These findings contributed to a Maxwell House decision to undertake additional marketing research and to convert its San Leandro, California, plant over four months ending late in May, 1963, in order to gain manufacturing and marketing experience with the no-key can. The San Leandro plant accounted for about 15 percent of the company's domestic production. The Hoboken, New Jersey, plant produced 40 percent of the total, and the remaining 45 percent was split about evenly between the plants in Jacksonville, Florida and Houston Texas.[1]

The following studies were started in February 1963, in an effort to obtain more evidence on how the keyless can would affect demand for Maxwell House ground coffee:

1. A survey to determine whether the favorable reaction to the keyless can voiced in November, 1962, by Stockton women had persisted

2. A survey to ascertain the awareness, attitudes, and purchasing behavior in regard to the keyless can among all ground-coffee users in Stockton

3. A survey of Muncie, Indiana, purchasers of Folger's coffee in the keyless can to learn of their reactions to the container and their subsequent purchasing plans and behavior

[1] Figures disguised.

4. A test in New York City to learn how users of ground coffee would evaluate the keyless can after continued use when they did not know the brand identity of the coffee contained in the can

The Stockton reinterviews showed that the new can continued to be regarded favorably, although the percentages of favorable responses ran somewhat lower in February than they had in November. Representative of the decline was the finding that the proportion of women who regarded the new can as better than the key can had dropped from 88 percent in November to 80 percent in February.

In general, the results obtained in the four different studies were similar and are indicated by findings of the survey of ground-coffee users in Stockton (see Exhibit 3–1 for summary).[2] Among the respondents who had tried the can, 95 percent liked at least something about it, 80 percent rated it better than the standard can on an overall basis, 83 percent said it was easier to open, and 84 percent rated it better on ease of reclosing. The percentages of respondents who had tried the new can and rated it either better or as good as the usual can ranged from 86 to 98 percent for each of several different qualities.

After examining the research findings, Maxwell House decided in July to convert all its plants to the keyless can. National conversion was achieved in October, 1963. Shortly afterward, most competing brands also converted. Maxwell House was the first brand to appear in the keyless can in markets accounting for 80 percent of its sales. Folger's was first with the new can in two markets, and Maxwell House introduced it simultaneously with one other brand in each of two markets. The length of time Maxwell House was the only brand in the keyless can varied considerably by markets but averaged from three to four months for the United States as a whole. At the end of this lead-time period, about two-thirds of the competitive ground-coffee volume had appeared in the new can.

Maxwell House's share of the ground-coffee market appeared to go up about half a share point several months after adoption of the keyless can. It was not known how much, if any, of the increase was a result of the container change, the concurrent introduction of a three-pound size, or other factors, but executives were inclined to give the new can some of the credit.

The keyless container cost an average of $.017 per can less than the key can, allowing for Maxwell House's mix of one- and two-pound sizes. Competition in the coffee industry typically had resulted in cost savings being passed on to the consumer in the form of special promotional deals or direct reductions in list prices. Adoption of the keyless can did not lead to an apparent decrease in the retail price of coffee, however, be-

[2] Exhibits follow Study Questions at end of chapter.

cause of the offsetting effect of an increase of two cents per pound in the cost of green coffee beans in the fall of 1963.

THE QUICK-STRIP CAN

Maxwell House executives considered the quick-strip container to be superior to both the keyless and the traditional key cans, combining their advantages. It had a metal cover sealed on the side just below the top with a flexible aluminum-coated plastic strip one-half-inch wide that easily could be pulled off for opening. For reclosure, it used its own cover, which had deeper side wells and provided a tighter fit, and thus would keep the contents fresher than the other two cans. The quick-strip can was the same size as the keyless can, which meant that costs of converting to it would be negligible.

The American Can Company had applied for several patents covering the principle of the tear-strip opening. Maxwell House had worked with the can company on the development of the quick-strip can. Although there was no exclusivity arrangement in regard to its use, Maxwell House executives were confident that they would have first rights to receive it for their needs until such time as American Can had developed sufficient capacity to supply the total industry. In view of the cost and time that probably would be necessary to develop such capacity, they believed that they would be able to offer the equivalent of eight months of their brand's national output in the new container without quick-strip-can competition.

If the can company followed its intended schedule of bringing in production capacity, Maxwell House would be able to average half of its output in the quick-strip can in the four months from mid-March to mid-July, achieving full output in the can on the latter date. It was assumed that all competing brands of ground coffee could be on the market in the new can by mid-March, 1965, if their managements elected to convert as soon as it was possible for them to do so. Once production capacity at the can company was available, a competitor probably could expect to have his entire output in the quick-strip can two months after placing his order. His conversion could be virtually instantaneous.

If Maxwell House were to adopt the quick-strip can, the hope would be that the new container would enable the brand to add to its share of market during the lead-time period and that at least some of the advantage in market position would be retained even though competitors also converted later. Executive opinions differed about the probable effect of the quick-strip can on Maxwell House sales, but they were within a range of possible average outcomes for the eight-month lead-time period of a loss of 1.5 percent of the ground-coffee market to a gain of 2.5 percent, assuming a retail-price increase of two cents per pound. One executive believed that the gain in share would be about twice as great if there were no price increase.

Although the executives were optimistic in regard to the quick-strip can, they recognized that its effect on demand was uncertain. They were extremely unwilling to take action that might lead to a decrease in market share. They feared that a decline in market position, once started, could not be contained. Because of the Maxwell House brand's large sales volume, a decrease could have a significant adverse effect on the General Foods Corporation's sales and profit figures at a time when great emphasis was being placed on growth and a decline in share was regarded as a signal of possible deterioration of a brand's consumer franchise.

COST CONSIDERATIONS

The quick-strip can was expected to cost $.007 per can more than the keyless container in the one-pound size and $.011 per can more in the two-pound size. This meant that Maxwell House's costs would go up an average of $.008 per can, allowing for the company's mix of the one- and two-pound sizes.

The cost picture prompted consideration of what pricing would be best for the quick-strip can. Prices might be left unchanged on the assumption that sales would increase more than enough to offset the higher container costs and result in higher profits before taxes. It was possible, however, that the quick-strip feature might prove so attractive that consumers would be willing to pay more to get it.

Changes in list prices of ground coffee of less than two cents per pound typically were not made because they were not large enough to result in immediate changes in retail shelf prices, especially in the large chain stores. Competitive pressures in the industry, however, typically would force increases in gross margin of less than two cents per pound realized by a roaster to be passed on in the form of special promotional deals. The trade discount for ground coffee was about 10 percent of the retail price.

MARKET POSITION

Maxwell House was the leading brand of ground coffee (drip, regular, and fine grind) in the United States, selling about 315 million pounds annually. The brand accounted for about 21.5 percent of all ground coffee sold, followed by Folger's, with about 15.1 percent (see Exhibit 3–2). If a packaging change should cause the brand to either lose or gain a 1-percent share of the national market, factory sales would be affected by about $9.3 million, assuming a selling price of $.65 per pound, and gross margin (manufacturer's sales minus cost of goods sold) would be affected by about $1.4 million.[3] Approximately 60 percent of the total pounds of Maxwell House ground coffee sold annually was packed in one-pound cans and 40 percent in two-pound cans.

Maxwell House was the leading brand of ground coffee on a national basis, but it was much stronger in the East than in the West. The opposite was true for Folger's. Maxwell House enjoyed 36 percent of the vacuum-packed ground-coffee market in its mideastern sales region, where Folger's share was 5.7 percent. In its west-central region, Maxwell House had a share of 7.8 percent compared with 31.8 percent for Folger's (see Exhibit 3–3).

Fifty-one percent of Maxwell House ground-coffee sales were made in the brand's eastern and mideastern sales districts, which accounted for only 5 percent of Folger's sales. Folger's was not sold in markets that accounted for about 40 percent of Maxwell House's sales. They included Boston, New York, Philadelphia, Syracuse, Washington, D.C., Youngstown, Charlotte, and Atlanta. The west-central and western regions contributed only 15 percent of the total sales of Maxwell House ground coffee, compared with 71 percent for Folger's.

The Maxwell House Division also roasted and marketed ground coffee under the Yuban brand and a decaffeinated ground coffee under the Sanka label. Yuban had about 1.9 percent and Sanka 1.5 percent of the ground-coffee market. The Maxwell House Division had maintained a position of leadership in the coffee industry for many years. Sales of its three brands in both ground and soluble form had grown to account for 35 percent of all coffee sold in the United States.

DECISION REQUIRED

In mid-November, the product manager wondered whether an immediate decision should be made to adopt or reject the quick-strip can, or whether the decision should be delayed to obtain evidence concerning its probable effect on demand. If it appeared advisable to try to obtain further information, decisions would have to be made concerning what research should be undertaken over what period of time.

Consideration was given to launching in mid-December market tests of the initial supply of 150,000 one-pound quick-strip cans in Stockton, California, and Muncie, Indiana, two towns that had been used by Folger's for sales tests of the keyless can. Maxwell House had considerable test-marketing and consumer-interviewing experience in both markets. Stockton represented the West, in which Maxwell House's market position was relatively weak. In Muncie, Maxwell House enjoyed

[3] Disguised figure.

a share of the ground-coffee market slightly higher than it held for the nation as a whole. Retail sales would be audited in twenty stores in Stockton and fifteen stores in Muncie, the outlets representing the great bulk of the coffee volume in both towns. Audits would be made before the test began, after the first week of the test, and then at two-week intervals. The audits would cost about $20 per store per visit. A majority of the families purchasing coffee in the one-pound size did so at least once every two weeks. The market research manager estimated that observation of three months of sales results would be required to ascertain how the change in cans was affecting Maxwell House's share of market.

In addition to the store audits, consumer interviews could be undertaken to learn how many people were aware of the new can, how many had tried it, and how they compared it with the keyless can. Telephone interviews of ground-coffee users would cost about $2 per interview, including tabulation of the responses. With more selective sampling requirements, the cost might go up to $4 per completed interview.

The company frequently had employed consumer-use tests, in which a new package was left in the home for trial. Later an interviewer would call to obtain the user's reactions and learn how he compared it with the type of package normally used. In the absence of unusual sampling requirements, the cost of a placement of a test package in the home and a call back would run about $12 per respondent. If two packages were placed in the home in successive weeks to be compared in use by the respondent, the two placements plus a call back would cost about $15 per respondent for a sample of ground-coffee users.

A new container could be test marketed in a Maxwell House sales district for virtually no additional research cost if the company relied on the reports it regularly received for each of its sales districts. Share-of-market figures by brand typically were received two weeks after the close of the two-month period to which they pertained.

In regard to the quick-strip can, it was the responsibility of the product manager in late 1963 to recommend a course of action. The division general manager would make the final decision.

STUDY QUESTIONS

ASSIGNMENT 1

1. Assume that you were the product manager for Maxwell House ground coffee, seeking to determine what course of action in regard to the quick-strip can you would recommend as of late 1963.
   a. What alternative courses of action would you have considered?
   b. What are the key questions that should be answered in the process of choosing among the alternatives?

c. Reasoning as best as you can on the basis of information presented in the chapter (without undertaking a formal prior analysis), what course of action would you have chosen to follow as of late 1963?

ASSIGNMENT 2

1. Structure decision tree branches for the following alternatives:

A_1 Maxwell House decides to convert completely to the quick-strip can as soon as possible.
A_2 Maxwell House stays with its present keyless can unless a major competitor (for example, Folger's) switches to the quick-strip can. If this should happen, Maxwell House would then decide whether to adopt the quick-strip can.

2. Prepare for conducting a formal Bayesian prior analysis of A_1 and A_2.
   a. What time period should be used?
   b. What measures would you use in comparing the alternatives?
   c. What kinds of calculations would you make?
   d. What inputs are needed for the calculations?
   e. Arrive at your subjective probabilities for use in the analysis, giving thought to a suitable procedure for doing so.
   f. What assumptions must be made for purposes of the analysis? (Keep a list of those you think should be used.)

3. Outline the main formulas needed for analyzing A_1.

ASSIGNMENT 3

1. Compute the expected value of A_1: Maxwell House decides to convert completely to the quick-strip can as soon as possible. To provide for more common ground for purposes of class discussion, it is suggested that the following assumptions and values be among those you use in your calculations:
   a. If a quick-strip can were to be adopted by Maxwell House, it would be offered either with no change in price or an increase in retail price of two cents per pound.
   b. Assume that an increase in retail price of two cents per pound also represents an increase of two cents per pound in Maxwell House's selling price (neglecting the 10 percent trade discount for purposes of simplification).
   c. Use a sixteen-month time period starting in mid-November, 1963, for calculating the expected change in Maxwell House profit before taxes.
   d. Assume that an increase in share of market has a positive value beyond the sixteen-month period of twice the gross margin for

the length of time the increase is in effect during the sixteen-month period. Assume that a drop in share of market has a negative value beyond the sixteen-month period of three times the gross margin for the length of time it is in effect during the sixteen-month period.

e. Use the following subjective probabilities and possible outcomes in Maxwell House share of market realized at the end of the sixteen-month period:

With price increase of two cents per pound:

Share of Market Range	Midpoint	Change in Share Points	Probability
19.0–20.99	20.0	−1.5	.25
21.0–21.99	21.5	0	.25
22.0–22.99	22.5	+1.0	.25
23.0–25.00	24.0	+2.5	.25
		Expected value of change:	+ .5

f. Assume that the change in share in effect at the end of the sixteen-month period is also the average change for the time Maxwell House is in the quick-strip can.

g. Assume that once 75 percent of the total ground-coffee volume is in the quick-strip can, the additional gross margin obtained from the price increase of two cents per pound will be passed on to the consumer in the form of special deals to meet promotional competition.

2. Without bothering to formally analyze A_2, how do you think its expected value would compare with the expected value of A_1?

ASSIGNMENT 4 (OPTIONAL)

1. Develop the formulas needed for analyzing A_2: Maxwell House stays with its present keyless can unless a major competitor (for example, Folger's) switches to the quick-strip can. If this should happen, Maxwell House would then decide whether to adopt the quick-strip can.

2. Compute the expected value of A_2.

ASSIGNMENT 5

1. Should Maxwell House attempt to obtain additional information before deciding whether to adopt the quick-strip can? Why or why not?

EXHIBIT 3–1

SELECTED FINDINGS FROM FEBRUARY, 1963, SURVEY OF GROUND-COFFEE USERS IN STOCKTON, CALIFORNIA

Interviews were completed with 399 users of ground coffee between February 16 and February 23, 1963. The sample was selected randomly from listings in the Stockton, California, telephone directory. Findings are summarized below.

1. Awareness and purchasing of new keyless can in the two months preceding the interview.

	Percent of All Respondents[a]		Percent of Respondents, by Usual Brand of Ground Coffee			
			Folger's		*Other*	
Had bought Folger's in new can	32		61		13	
Aware of new can but had not purchased	40		33		44	
Aware that new can is Folger's		26		25		26
Believe new can to be another brand		6				8
Don't know brand		8		8		10
Not aware of new can	28		6		43	
Total	100		100		100	
Number of respondents	(399)		(157)		(242)	

[a] Fifty-six percent of all respondents had bought Folger's Coffee in either the keyless or the key can in the two months before the interview. The corresponding figures for the women whose usual brand was Folger's and for the women who usually bought another brand were 98 and 29 percent, respectively.

2. Question: "Has the Folger's Coffee you bought been in a new kind of can or in the usual kind of can that most ground coffees are sold in?"

	Percent of All Respondents	Percent of Respondents, by Usual Brand of Ground Coffee	
		Folger's	*Other*
In new can only	35	35	33
In both new and regular can	22	27	13
Total any new can	57	62	46
In regular can only	43	38	54
Total	100	100	100
Number of respondents who had bought Folger's	(223)	(153)	(70)

EXHIBIT 3–1 (Continued)

3. Question (if bought Folger's in new can): "What size was this new can—how much coffee did it contain? How often have you bought Folger's in this new kind of can during the past two months?"

Size of Can and Times Purchased	Percent of All Respondents	Percent of Respondents, by Usual Brand of Ground Coffee	
		Folger's	*Other*
Size of can purchased [b]			
1 lb	61	58	69
2 lb	2	3	
3 lb	36	38	31
Don't know	14	18	3
Number of times purchased			
1	40	37	50
2	18	17	22
3	18	19	16
4	12	13	9
5 or more	12	14	9
Total	100	100	100
Number of respondents who bought new can	(127)	(95)	(32)

[b] Percentages add up to more than 100 because some respondents had purchased more than one size.

4. Question: "Would you say that the new coffee can was, generally speaking, better or worse or about the same as the usual kind of coffee can? Is that a good deal better (worse) or only a little better (worse)?"

	Percent of All Respondents		Percent of Respondents, by Usual Brand of Ground Coffee			
			Folger's		*Other*	
Better	80		79		81	
A good deal better		68		66		72
Only a little better		12		13		9
About the same	12		13		10	
Worse [c]	7		7		6	
Only a little worse		4		3		6
A good deal worse		3		4		
Don't know	2		1		3	
Total	100		100		100	
Number of respondents who bought new can	(127)		(95)		(32)	

[c] Of the nine women who considered the new can worse than the usual container, six disliked use of the can opener or found the can hard to open and two said it was hard to close or hard to get the coffee out.

EXHIBIT 3–1 (Continued)

5. Question: "What, if anything, did you particularly like about Folger's new can?"

Response			Percent of All Respondents
Liked something about the new can			95
Ease of opening		45	
Easy to open; quicker; no spilling	26		
No trouble with key; like can opener	15		
Safer; no sharp edges; no danger of cuts	13		
Plastic lid		26	
New plastic top; lid	13		
Coffee stays fresh; seals better	9		
Easy to reclose	7		
Reuse as canister		56 [d]	
Can reuse for storing as canister	39		
Pretty colors; attractive decals	24		
Can use as a coffee canister	6		
Size and shape			
Stores well; compact	13		
Easy to handle; hold; grip	9		
Like the size (no mention of pounds)	6		
Like the 3-lb size		3	
Just another can		3	
No specific reason		2	
Do not like anything about new can			5
			100
(Number of respondents who bought new can)		(127)	

[d] Twenty-five percent referred specifically to the 3-pound can; 17 percent mentioned reuse of the 3-pound container.

EXHIBIT 3–1 (Continued)

6. Questions: "Now for each of the following, would you tell me whether the Folger's Coffee in the new type of can was as good as what you used to get in the usual kind of can or better or worse? Compared to the usual type of coffee can, is it more difficult, about the same, or less difficult to get the last few spoonfuls out of the can?"

Evaluation of Coffee and New Can	Percent of All Respondents (*N* = *127*)	Number of Respondents, by Usual Brand of Ground Coffee: *Folger's* (*N* = *95*)	*Other* (*N* = *32*)
Evaluation of Coffee in New Cans			
Freshness			
Better than in usual can	28	24	41
As good as in usual can	70	75	56
Worse than in usual can			
Don't know	2	1	3
Aroma			
Better than in usual can	24	23	28
As good as in usual can	71	73	66
Worse than in usual can	1	1	
Don't know	4	3	6
Flavor			
Better than in usual can	24	20	34
As good as in usual can	73	79	57
Worse than in usual can			
Don't know	3	1	9
Evaluation of New Can			
Ease of opening can			
Better than usual can	83	83	84
As good as usual can	7	8	3
Worse than usual can	9	9	6
Don't know	1		6
Ease of reclosing can			
Better than usual can	84	86	78
As good as usual can	9	8	12
Worse than usual can	4	4	3
Don't know	3	2	7
Ease of removing last spoonfuls			
Better (less difficult)	28	29	25
About the same	58	57	59
Worse (more difficult)	12	12	12
Don't know	2	2	4

EXHIBIT 3–1 (Concluded)

7. Question: "Actually, of course, I am talking about the new Folger's can, which is opened with a can opener instead of the usual little key, and which has a slip-on plastic lid, which you use to cap the can after it has been opened. Would you want your own regular brand to come in such a can?" (Asked of all those who had not bought Folger's in the new type of can.)

Attitudes Toward New Can	Percent of All Respondents		Percent of Respondents, by Usual Brand of Ground Coffee: *Folger's*		Percent of Respondents, by Usual Brand of Ground Coffee: *Other*	
Would want usual brand to come in such a can	55		75		49	
Have seen or heard of new can		38		67		29
Have not seen or heard of new can		17		8		20
Would not want usual brand to come in such a can	37		20		42	
Have seen or heard of new can		15		13		15
Have not seen or heard of new can		22		7		27
Don't know; no answer	8		5		9	
Total	100		100		100	
Number of respondents who had not bought Folger's in new can	(272)		(62)		(210)	

8. Question (if did not want usual brand to come in the new one): "Why is that?"

Reasons for Preferring Old Can		Percent of All Respondents
Prefer present can		44
No problem with key; like the key	24	
Satisfied with way it is now	17	
If they change the can, might change flavor	4	
Can does not matter		33
Don't see where it would matter	16	
Take it out of can and put in canister	9	
Buy coffee, not containers	9	
All other reasons		14
Don't know; no answer		9
Total		100
Number of respondents who did not want usual brand to come in the new can	(100)	

Source: Marketing Research Department, Maxwell House Division, General Foods Corporation.

EXHIBIT 3-2

ESTIMATED PERCENTAGE SHARE OF GROUND-COFFEE SALES FOR MAXWELL HOUSE, FOLGER'S, SANKA, AND YUBAN BRANDS IN THE UNITED STATES, 1953 TO 1963

Brand	Estimated Percentage							
	1953	1955	1957	1959	1960	1961	1962	1963[a]
Maxwell House	15.6	16.3	16.7	19.1	20.8	21.6	21.4	21.5
Folger's					14.6	14.8	14.7	15.1
Sanka	1.1	0.8	1.0	1.1	1.2	1.3	1.5	1.8
Yuban			0.7	0.8	1.0	1.8	1.9	1.7

[a] Estimates for late 1963.

SOURCE: Marketing Research Department, Maxwell House Division, General Foods Corporation.

EXHIBIT 3-3

ESTIMATED PERCENTAGE SHARES OF VACUUM-PACKED GROUND-COFFEE SALES FOR SELECTED BRANDS, BY MAXWELL HOUSE SALES REGIONS, FIRST EIGHT MONTHS, 1963

Region	Estimated Percentage				
	Maxwell House	*Yuban*	*Folger's*	*Chase & Sanborn*	*Hills Bros.*
Eastern (Boston, New York, Philadelphia, Syracuse)	32.4	2.4		9.3	0.4
Mideastern (Washington, Youngstown, Cincinnati, Louisville)	36.0	1.2	5.7	11.2	3.6
Southern (Charlotte, Atlanta, Jacksonville, Memphis, New Orleans)	32.7	0.7	10.6	6.9	0.4
Central (Detroit, Indianapolis, Chicago, Milwaukee, St. Louis)	15.7	0.5	11.3	7.5	25.0
West Central (Minneapolis, Omaha, Kansas City, Dallas, Houston)	7.8	0.1	31.8	1.6	5.8
Western (Portland, San Francisco, Los Angeles, Denver, Phoenix)	8.7	4.9	26.9	3.6	16.5

SOURCE: Marketing Research Department, Maxwell House Division, General Foods Corporation.

CHAPTER 4

PLANNING AND CONDUCTING THE PRIOR ANALYSIS

In this chapter, several alternative courses of action for dealing with the packaging-decision problem presented in Chapter 3 are identified for special attention. The approach to their evaluation is described and two prior analyses, one with and one without the use of the computer, are presented in detail. Alternatives that involve seeking additional information are considered in Chapter 5, which focuses on preposterior analysis.

ALTERNATIVE COURSES OF ACTION

The number of possible alternatives is large, but they are of the following kinds:

1. Decide now to convert to the quick-strip can (must specify timing and extent of conversion and price to be charged).

2. Decide now to stay with the keyless can until provoked by some specified event (such as adoption of quick-strip can by a major competitor), then reconsider.

3. Decide now to order research to obtain additional information before deciding whether to order equipment needed to convert to the quick-strip can.

4. Place the order now for the equipment needed for conversion to the quick-strip can, and order research to obtain additional information to be used in deciding whether to go ahead with conversion.

Attention will be focused on several alternatives representing the above categories. These alternatives have been selected because of their value for illustrating important aspects of applying decision theory under realistic operating conditions.

APPROACH TO THE ANALYSIS

The approach to handling the packaging-decision problem consists first of diagramming it in the form of a decision tree in which the alternatives under consideration are represented by primary branches (see Exhibit 4–1). The latter will then be developed to provide the structure for the analysis in which alternatives can be examined one by one to determine how they would affect Maxwell House Division profits.

The following alternatives were selected for attention for illustrative purposes:

A_1 Convert completely to the quick-strip can as soon as possible with either (*a*) no change in price or (*b*) a price increase of two cents per pound.

A_2 Stay with the keyless can unless a major competitor (for example, Folger's) switches to the quick-strip can. If this should happen, Maxwell House then would decide whether to adopt the quick-strip can.

A_3 Conduct a three-month sales test in Muncie and Stockton of the one-pound quick-strip can with a price increase of two cents per pound; then decide whether to adopt or reject the new can. Equipment needed for conversion can be ordered either (*a*) immediately or (*b*) after reviewing the results of the sales test.

A_4 Conduct the sales test mentioned in A_3, consumer interviews in the test markets, and an in-home use test; then decide whether to adopt or reject the quick-strip can. Equipment can be ordered either (*a*) immediately or (*b*) after reviewing the results of the research.

These alternatives are consistent with information presented in Chapter 3. Consideration of no price change other than an increase of two cents per pound reflects the judgment that the consumer would be unwilling to pay more than that. In addition, it recognizes the higher cost of the new container and executive opinion that a lesser change in price would not be operationally feasible. The research alternatives reflect the pressure to get evidence as soon as possible so that the assumed lead time over competition with respect to use of the quick-strip can would not be sacrificed.

Alternatives A_1 and A_2 will be discussed in this chapter; alternatives A_3 and A_4 will be discussed in Chapter 5. Analysis of the alternatives will take into account their consequences in terms of costs, share of market, and profits. Inasmuch as the total ground-coffee market was not expected to change much in the near future or to be affected by the proposed package change, share of market was the measure of sales outcome about which there was uncertainty.

A DECISION TREE BRANCH FOR A_1

In building a decision tree, it is necessary to decide how many factors will be represented in the structure of the tree and how many will be considered only in work-sheet computations. To implement the alternative of completely converting to the quick-strip can as soon as possible, a choice must be made between price alternatives. These are built into the tree structure. Also incorporated are different consumer reactions measured in terms of changes in share of market and their financial consequences. The uncertainty of what competitors will do about the quick-strip can is not represented explicitly in the tree (although it could have been) but will be taken into account in the work sheets and will be reflected in the end values of the tree. The decision tree branch for A_1, then, is kept relatively simple (see Figure 4–1).

FIGURE 4-1

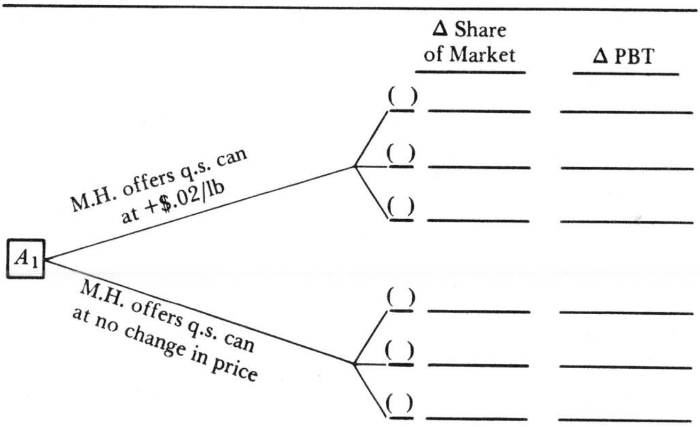

A_1

M.H. offers q.s. can at +$.02/lb

M.H. offers q.s. can at no change in price

	Δ Share of Market	Δ PBT
( )		
( )		
( )		
( )		
( )		
( )		

A DECISION TREE BRANCH FOR A_2

The tree structure for A_2 is more complex because it provides for a gaming situation in regard to the cost of the quick-strip can (see Figure 4–2).

Under A_2, Maxwell House decides not to adopt the quick-strip can now. The can company, therefore, probably would try to sell the can to another coffee roaster. One major branch of the tree represents what could happen if a major competitor adopted the quick-strip can. The competitor could offer coffee in the new can at a price increase of two cents per pound or without changing the price. Maxwell House then could observe the share-of-market results from audits of retail store sales. If the competitor's offering were well enough received (a success according

FIGURE 4-2

DECISION TREE BRANCH FOR A_2

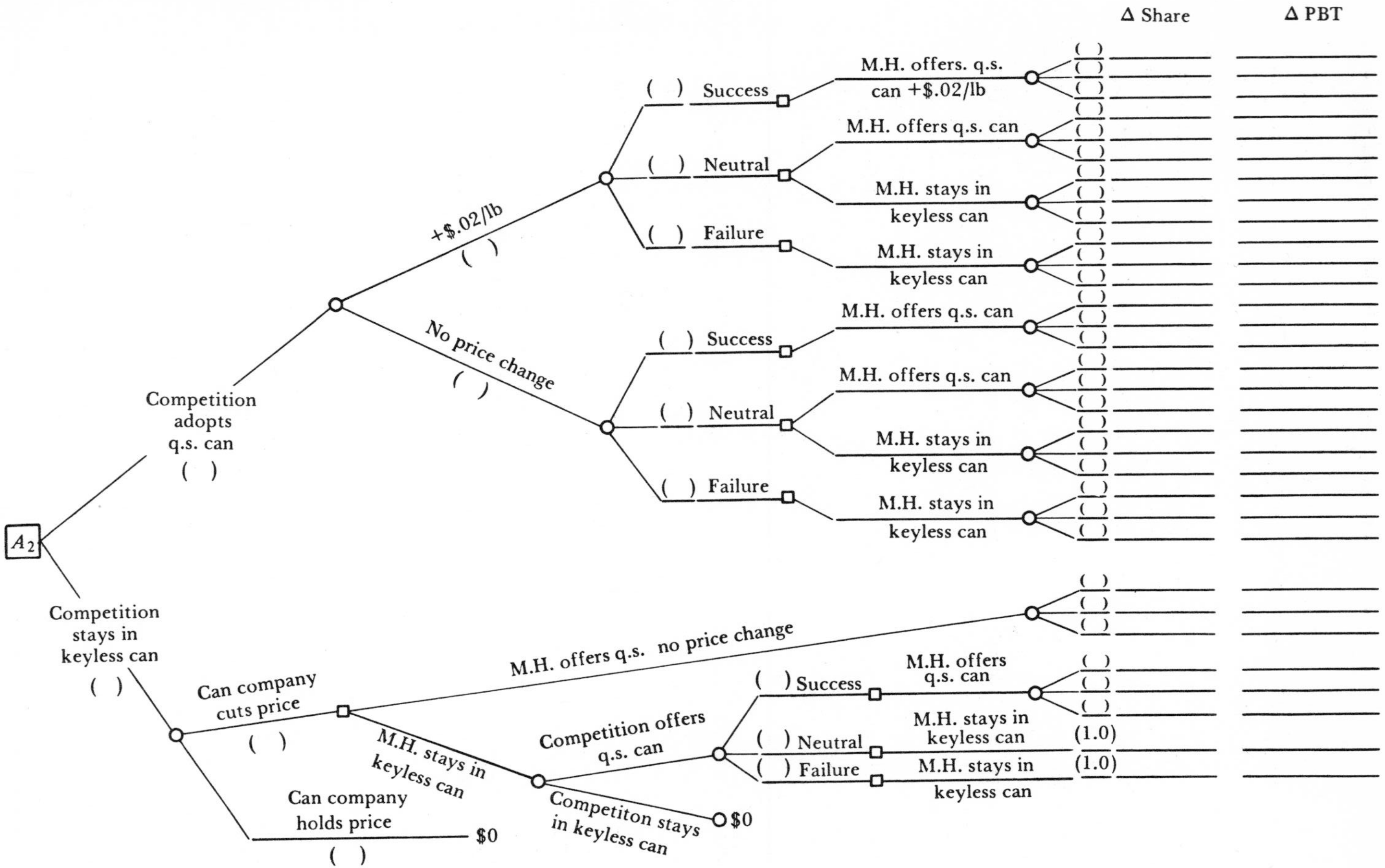

to predetermined criteria) Maxwell House would adopt the can. The tree allows for three values representing the range of possible effects of this action on Maxwell Houses's share of market and profit before taxes. If results of the competitor's quick-strip-can offering clearly were unfavorable, Maxwell House would remain in the keyless can. If the competitor's quick-strip-can offering were less clear-cut (neutral), the tree provides for considering the effects of both adoption and rejection of the new can by Maxwell House.

In the event that competition did not decide to adopt the quick-strip can (see lower branch), the can company might then offer the can to Maxwell House at a reduced price. The tree provides for considering the possible outcomes of adopting the new can on this basis. If Maxwell House refused the new offer, competition might choose to adopt the quick-strip can at the reduced price. If it did, Maxwell House could observe the share-of-market data to note the effect and act accordingly. If competition decided to remain in the keyless can, that might end the matter as indicated by the tree as it is shown in Figure 4–2. The tree could be extended to provide for another price-reduction attempt by the can company if such an attempt were regarded as likely.

PRIOR ANALYSIS OF A_1 WITHOUT USING THE COMPUTER

CHOOSE TIME PERIOD. The time period should be long enough to permit consideration of the full effect of adoption of the quick-strip can on Maxwell House's share of market, allowing for the influences of consumer reaction to the can and the effect of competitor's actions. One possible approach is to choose a time period long enough so that change in Maxwell House share of market due to adoption of the quick-strip can by Maxwell House and its competitors will have stopped. In this connection, it is useful to build a timetable such as the following:

		Month	
Mid-November,	1963	0	Time at end of decision problem described in Chapter 3.
Mid-December,	1963	1	Maxwell House decides to convert as soon as possible.
Mid-January,	1964	2	
		3	
Mid-March,	1964	4	Maxwell House output in quick-strip can starts.
		5	
		6	
		7	
Mid-July,	1964	8	Maxwell House reaches full output in quick-strip can.
		9	
		10	
		11	
Mid-November,	1964	12	Some competition could enter with quick-strip can.
		13	

		Month	
Mid-January,	1965	14	
		15	
Mid-March,	1965	16	All competitors could be in quick-strip can if their managements elected as soon as possible to convert.

In accordance with information presented in Chapter 3, it is assumed that Maxwell House would realize the equivalent of two full months of output in the quick-strip can during the four buildup months that start with mid-March, 1964, and that Maxwell House would enjoy the equivalent of eight full output months in the quick-strip can before encountering quick-strip-can competition.

All competitors could be in the quick-strip can if they decided as soon as possible to convert. It is not known, of course, how many would decide to adopt the quick-strip can or how soon. The timetable above is based on the assumption that competitive quick-strip-can coffee volume would be large enough so that change in Maxwell House's share of market due to the quick-strip can would have stopped by the end of the sixteenth month. If Maxwell House were to remain in the quick-strip can after introducing it in accordance with the above schedule, it would realize the equivalent of ten full months of quick-strip-can output during the sixteen-month period. The handling of the analysis to be described uses a sixteen-month period initially. Valuing any change in Maxwell House's share that still may be in effect at the end of the sixteen months is done in another step, which will be described later.

ASSIGN PROBABILITIES TO POSSIBLE CHANGES IN SHARE OF MARKET. In arriving at subjective probabilities for changes in Maxwell House's share of market that might result from offering the quick-strip can, the decision maker should do the following:

1. Appraise available information on effects of changes in package and price. (Chapter 3 reported the company's recent experience in changing to the keyless can.)

2. Judge the nature of the proposed change to determine what effects it might have on consumer demand.

3. Establish ranges of possible outcomes of adopting the quick-strip can for the combination of events represented by branches of the decision tree.

4. Break down the ranges and assign probabilities to the subdivisions.

There is no one way to arrive at the subjective probabilities. A systematic approach is desirable, however, because of the importance of the values to the analysis. One approach is described below. It involves confronting the decision maker with a line on which there are points representing the possible share-of-market outcomes for Maxwell House of adopting the quick-strip can.

18.5 19.5 20.5 21.5 22.5 23.5 24.5

The decision maker is asked to assume that Maxwell House offers its ground coffee in the quick-strip can at a price increase of two cents per pound and that it has no quick-strip-can competition for ten months. He then is asked to assign probabilities to what he regards to be the possible outcomes in terms of average share of market for Maxwell House for the ten months. He is asked to follow this procedure:

1. Draw two vertical lines through the horizontal line (above) so that the space between them represents the range of possible average share-of-market outcomes.

2. Draw a line at the point within the range at which it is just as likely that the actual result would be above that value as it would be below. (This step divides the range into two segments of equal subjective probability of occurrence.)

3. Now, assume that the actual result would be between the lowest point of the range and the vertical line just drawn. Divide this segment by drawing a vertical line at the point at which it is just as likely that the actual result would be above that value as it would be below. (This step divides one of the segments of .5 probability into two segments of equal subjective probability of occurrence.)

4. The above step now is repeated for the other segment of .5 probability. The range now has been divided into quadrants of equal subjective probability of occurrence. Midpoints of the quadrants are calculated for use in computations.

Results of following the above procedure might look like this depending, of course, on the decision maker's judgment.

Share of Market Range	Midpoint	Change in Share Points	Probability
19.0–20.99	20.0	−1.5	.25
21.0–21.99	21.5	0	.25
22.0–22.99	22.5	+1.0	.25
23.0–25.00	24.0	+2.5	.25
		Expected value of change:	+.5

The above steps could be repeated to arrive at subjective probabilities appropriate to an assumption of no change in price.

Other procedures, of course, can be used to arrive at the subjective probabilities. Quadrants of equal subjective probability are not essential to the conduct of the analysis. They were the product of the above procedure, which was chosen not because it produced quadrants but because it seemed to represent a feasible way of imposing desirable rigor on the decision maker.

IDENTIFY KINDS OF CALCULATIONS NEEDED. The analysis for the sixteen-month time period will entail calculations of changes in the following values for different circumstances and for different periods of time in which Maxwell House may be in the quick-strip can:

1. *Normal gross margin* refers to the gross margin (manufacturer's dollar sales minus cost of goods sold) Maxwell House would realize, using the same gross margin rate that pertained in the past. It does not allow for the effect of a price change.

2. Revenue due to price change

3. Can costs

4. Promotion costs

Values for the first three factors can be calculated at the outset on a per-share-point per-month basis as follows:

1. Normal gross margin/share point/month $= \dfrac{\$1{,}400{,}000}{12 \text{ months}} = \$116{,}667$

2. Change in revenue/share point/month due to price increase =

$$\frac{\$0.02/\text{lb} \times 315{,}000{,}000 \text{ lb/yr}}{21.5 \text{ share points} \times 12 \text{ months}} = \$24{,}419$$

3. Change in can cost/share point/month =

$$\frac{252{,}000{,}000 \text{ cans/yr} \times \$0.008/\text{can}}{12 \text{ months} \times 21.5 \text{ share points}} = \$7{,}814$$

(Cans/yr = (315,000,000 × .6) + (315,000,000 × .4) = 252,000,000)

VALUE CHANGE IN SHARE PRESENT AT END OF SIXTEEN-MONTH PERIOD. Expected monetary value of the course of action computed on the basis of the sixteen-month period does not value any change in share still in effect at the end of the sixteenth month for its economic consequences beyond that time. How the post-sixteen-months value of a change in share might be taken into account will be examined in this section.

The approach followed in this case was to ask the decision maker how much he would be willing to pay for an increase of 1 share point that he would be guaranteed he would hold for ten months. He then was asked similar questions in regard to increases of 2 and 3 points. For a decrease in share, he was asked how much he would have to be paid to lead him to accept a reduction of 1 share point that he knew would be in effect for at least ten months. He was asked similar questions in regard to greater decreases in share.

For purposes of illustrative computations, we shall assume that the decision maker's answers to these questions showed that he placed the future value of an increase in share, after it had been held for ten months, at twice the normal gross margin for a ten-month period. We shall also assume that his answers showed that he placed the future value of a decrease in share, after it had been in effect for ten months, at minus three times the normal gross margin for a ten-month period.

DEVELOP FORMULAS FOR COMPUTATIONS. A general statement of computations needed for arriving at the change in profits before taxes (PBT) resulting from offering the quick-strip can follows:

$$\Delta\,\text{PBT} = \Delta\,\text{PBT for 16-month period} + \Delta\,\text{PBT for time beyond 16-month period}$$

$$\begin{array}{l}\Delta\,\text{PBT for}\\ \text{16-month}\\ \text{period}\end{array} = \left[\begin{array}{l}\Delta\text{ share}\\ \text{points for}\\ \text{months of}\\ \text{q.s. output}\end{array} \times \begin{array}{l}\text{normal gross}\\ \text{margin/}\\ \text{share point/}\\ \text{month}\end{array} \times \begin{array}{l}\text{months}\\ \text{of q.s.}\\ \text{can}\\ \text{output}\end{array}\right]$$

$$+ \left[\begin{array}{l}\text{share}\\ \text{points for}\\ \text{months of}\\ \text{q.s. output}\end{array} \times \begin{array}{l}\Delta\text{ revenue/}\\ \text{share point/}\\ \text{month due}\\ \text{to price}\\ \text{increase}\end{array} \times \begin{array}{l}\text{months}\\ \text{of q.s.}\\ \text{can}\\ \text{output}\end{array}\right]$$

$$- \left[\begin{array}{l}\text{share}\\ \text{points for}\\ \text{months of}\\ \text{q.s. output}\end{array} \times \begin{array}{l}\Delta\text{ can costs/}\\ \text{share point/}\\ \text{month}\end{array} \times \begin{array}{l}\text{months}\\ \text{of q.s.}\\ \text{can}\\ \text{output}\end{array}\right] - \begin{array}{l}\text{added}\\ \text{promotional}\\ \text{costs}\end{array}$$

$$\begin{array}{l}\Delta\,\text{PBT}\\ \text{for time}\\ \text{beyond}\\ \text{16 months}\end{array} = \begin{array}{l}\Delta\text{ share}\\ \text{points for}\\ \text{time in q.s.}\\ \text{can during}\\ \text{16 months}\end{array} \times \begin{array}{l}\text{normal gross}\\ \text{margin/}\\ \text{share point/}\\ \text{month}\end{array} \times \begin{array}{l}\text{months}\\ \text{of q.s.}\\ \text{output}\\ \text{in 16}\\ \text{months}\end{array} \times$$

COMPUTE EXPECTED CHANGE IN PROFITS FOR A_1. The decision tree branch for A_1 shown in Figure 4–3 illustrates for each price alternative: (1) the expected changes in market-share points, (2) the consequences of each change in market share in terms of expected change in profits before taxes, and (3) the expected value of change in profits before taxes of the price alternative. The alternative of no change in price is rejected because its expected value (−$235,000) is lower than that for a price increase of two cents per pound ($5,058,000), because its exposure to loss is greater, and because its exposure to gain is less. These results are the product of detailed computations shown in Exhibit 4–2. (See the decision tree in Figure 4–3 for a definition of states.) An analysis using

FIGURE 4–3

DECISION TREE BRANCH FOR A_1

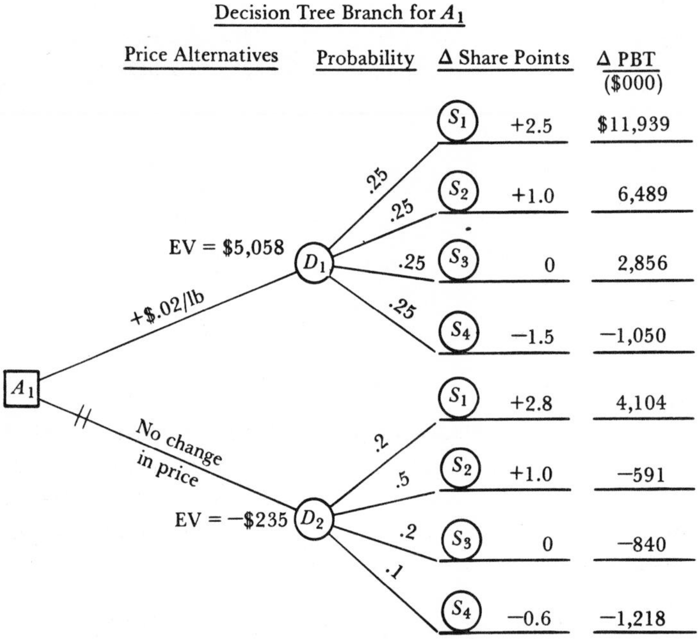

different assumptions, of course, would produce different expected values.

If the decision maker thought that the price alternative of no change would not lead to an increase of more than 1.5 share points, he might have omitted the alternative from the prior analysis. The argument is outlined as follows:

1. Gain in share points needed to break even (i.e., to meet the higher can costs) =

$$\frac{\text{increase in can costs}}{\text{gain in gross margin/share point from increased volume}}$$

2. Increase in can costs for 1 yr with quick-strip can = \$.008/can × 252,000,000 cans = \$2,016,000

3. Net gain in annual gross margin from 1 additional share profit if quick-strip can is used:

Gross margin/yr/share point	=	\$1,400,000
Added can cost/yr/share point	=	93,767
		\$1,306,233

$$\left(\text{Added can cost/yr/point} = \frac{\$.008 \times 252{,}000{,}000 \text{ cans}}{21.5 \text{ share points}}\right)$$

4. Gain in share points needed to break even $= \dfrac{\$2{,}016{,}000}{\$1{,}306{,}233} = 1.47$ share points

5. Expected value of change in share points if quick-strip can is used at no price increase (using subjective probabilities given below):

Change in Share Points	Probability	Expected Value
−0.6	.1	− .06
0	.2	.00
+1.0	.5	.50
+2.8	.2	.56
		+1.00

6. The above computations show that the expected value of the increase in share of market is substantially lower than that needed to break even. On the basis of the subjective probabilities used in this analysis, offering the quick-strip can at no change in price is not a good bet for Maxwell House.

PRIOR ANALYSIS OF A_1 USING THE COMPUTER WITH TIME-SHARING SYSTEM

THE MODEL. Subsequent to the development of the analytical plan followed in the preceding section, a computer program was developed for the task. It was written in Fortran IV language for use on an IBM 360/67, the inputs and print-outs to be made by means of a teletype console connected with the computer by telephone line.

Using the computer, of course, relieves human hands of the computational burden, making it possible to focus attention on the important function of developing input data. The structure of the computer model is indicated by the input data sheet that appears in Exhibit 4–3. Certain features that represent improvements over the model described in the preceding section will be discussed here.

In the model for analysis by hand, a given expected outcome for the quick-strip can was expressed as one figure, an average change in share of market for a specified number of months within the chosen sixteen-month period. In the model for computer analysis, an expected share-of-market outcome is expressed as a profile consisting of three connected straight lines which recognize that the share of market may change over time. The profile starts from 21.5 share points at the time quick-strip output begins. To form the remainder of the profile, the decision maker is asked for the following inputs:

1. Market share at point of maximum change from 21.5

2. The point in time at which the maximum change in market share will be attained

3. The number of months the maximum change in market share is expected to be held

4. The point in time when the change in market share due to the quick-strip can and its accompanying price will become completely dissipated (or beyond which management is unwilling to value the change)

FIGURE 4-4

ILLUSTRATION OF MARKET-SHARE PROFILES

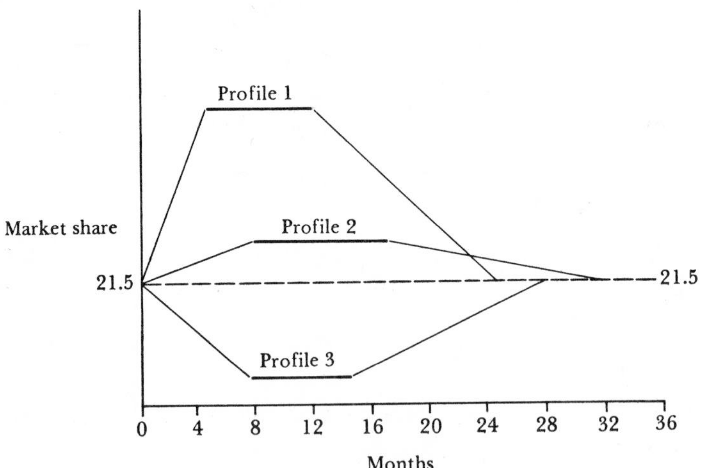

The range of expected outcomes of a given course of action is represented by several market-share profiles, each of which has an assigned probability of occurrence. (In Figure 4–4, the range of expected outcomes is represented by a set of three profiles.) The computer calculates the expected change in profits before taxes for each profile. It then multiplies each of these figures by its probability and sums the products to get the expected value of the change in profits for the alternative.

The use of market-share profiles makes it easier for the decision maker to visualize sales volumes that will change over time in response to competitive actions and other factors. Inasmuch as each profile has its own time period that extends for as long as management is willing to place a value on a change in market share, an adjustment for any future value of a change still in effect at the end of a predetermined time period of analysis (such as the adjustment that had to be made in the earlier analysis) is not necessary.

If the extent and duration of a change in share of market would cause the decision maker to decide to reconvert to the keyless can, he can so indicate. He is then asked to specify the time at which he would make the decision to reconvert, the additional cost that he expects to be associated with reconverting and attempting to regain lost market share, and the number of months that he feels will be required to regain the original 21.5 share-of-market position.

PRIOR ANALYSIS OUTPUT. A sample print-out of a prior analysis of the action of offering the quick-strip can at a price increase of two cents per pound is shown in Exhibit 4–4. It includes the inputs used, the expected change in profits before taxes for the course of action, and the following information for each market-share profile:

1. Change in total gross margin due to change in market share, exclusive of the effect of any change in price

2. Change in total gross margin due to change in price, assuming the new price is maintained for the expected life of the market-share profile

3. Change in can costs

4. Incremental gross margin due to price increase that subsequently is lost because of competitive promotional and pricing activities

5. Reconversion costs

6. Change in profits before taxes (the net result of the five items listed above)

The expected change in profits before taxes for offering the quick-strip can at a price increase of two cents per pound was found to be $4,423,410 in the example represented in Exhibit 4–4 (computed by multiplying the expected change in profits for each profile by the profile's probability and summing the products). The exposure to gain and loss is indicated by the expected change in profits for each profile, ranging from a loss of $1,253,081 with a .3 probability to a gain of $8,116,587 with a .4 probability. (*Note:* the inputs for the computer analysis differed somewhat from those used in the prior analysis in which the computer was not used.)

Results of the prior analysis by computer of A_1 with no change in price are not presented here, but they compared unfavorably with those for A_1 with a price increase just as they did in the analysis done without using the computer.

PRIOR ANALYSIS OF A_2

A prior analysis of A_2, which would include steps similar to those followed for A_1, is not presented here because of space limitations. If the decision maker's prior probabilities for the quick-strip can were optimistic (as in the examples described in this chapter), formal analysis of A_2 would not be necessary. It could be seen by inspection that the expected change in profits for A_2 would be less than for A_1 with the price increase of two cents per pound. The reason is that A_2 involves a delay in realizing what is expected to be a gain from adopting the new can. Exposure to gain under A_2 would be less than under A_1, and it is questionable whether exposure to loss would be any less. If it was believed that use of the quick-strip can would increase sales of competing coffees at the expense of the Maxwell House brand for the time the latter is not in the new can, the expected value of change in profits for A_2 could be negative.

STUDY QUESTIONS

1. Compare the computer model for the prior analysis of A_1 with the model used in the analysis done without the computer. Which model do you prefer? Why?

2. What procedure do you prefer for arriving at subjective probabilities? Why?

3. How did the results reported in the chapter compare with those of your own prior analysis (see Chapter 3, Assignment 3)? How do you account for the main differences?

4. If Maxwell House were going to seek additional information for use in deciding what to do about the quick-strip can, what research plan do you think would be most appropriate? Why?

EXHIBIT 4–1

DECISION TREE OF ALTERNATIVES CONSIDERED IN CHAPTERS 4 AND 5

A_1 Convert completely to quick-strip can as soon as possible with (*a*) no change in price or (*b*) increase of two cents per pound.

A_2 Stay with keyless can unless a major competitor switches to quick-strip can; then reconsider.

A_3 Conduct three-month sales test in Muncie and Stockton; then decide. Order equipment needed for conversion either (*a*) immediately or (*b*) after reviewing sales-test results.

A_4 Conduct sales test, consumer interviews, and in-home use test; then decide. Order equipment needed for conversion either (*a*) immediately or (*b*) after reviewing research results.

EXHIBIT 4–2

COMPUTATIONS FOR PRIOR ANALYSIS OF A_1

(In Thousands of Dollars)

	D_1S_1	D_1S_2	D_1S_3	D_1S_4	D_2S_1	D_2S_2	D_2S_3	D_2S_4
Avg. M.H. share for time in q.s. can	24.0	22.5	21.5	20.38	24.3	22.5	21.5	21.05
Δ M.H. share points for time in q.s. can	+2.5	+1.0	0	−1.12	+2.8	+1.0	0	−.45
Δ PBT for 16-Month Period								
Δ normal gross margin [a]	2,917	1,167	0	−1,050	3,267	1,167	0	−560
Increase in revenue due to price increase [b]	5,861	5,494	5,250	1,991				
Line 3 + line 4	8,778	6,661	5,250	941	3,267	1,167	0	−560
Increase in can costs [c]	1,875	1,758	1,680	637	1,899	1,758	840	658
Increase in promotion costs [d]	797	747	714	1,354				
Line 6 + line 7	2,672	2,505	2,394	1,991	1,899	1,758	840	658
Δ PBT for 16-month period [e]	6,106	4,156	2,856	−1,050	1,368	−591	−840	−1,218
Value of Share Beyond 16-Month Period [f]	5,833	2,333	0	0	2,736	0	0	0
Adjusted monetary value [g]	11,939	6,489	2,856	−1,050	4,104	−591	−840	−1,218
Probability	.25	.25	.25	.25	.20	.50	.20	.10
Expected value [h]	2,985	1,622	714	−263	821	−296	−168	−122
	Expected value for D_1 = 5,058				Expected value for D_2 = 235			

Explanatory Notes for All States Except D_1S_4, D_2S_2, D_2S_3, and D_2S_4

[a] Δ Normal gross margin = normal gross margin/month/share point ($116,667) × months of quick-strip output (10) × Δ M.H. share points

[b] Δ Revenue due to Δ price = average M.H. share points for months of quick-strip output × revenue/share point/month due to price increase × number of months in quick-strip can

[c] Δ Can costs = Δ can costs/share point/month ($7,814) × number of months of quick-strip can output for Maxwell House (10) × M.H. share points

[d] Δ Promotion costs = (Δ revenue/share point/month due to price increase − Δ can costs/share point/month) × M.H. share points × number of months in which more than 75 percent of industry ground-coffee volume is in quick-strip can) = ($24,419 − $7,814) × (M.H. share points × 2)

[e] Line 5 − line 8

[f] Value of Δ share beyond 16-month period = Δ M.H. share points for months of quick-strip-can output during 16-month period × normal gross margin/share point/month ($116,667) × number of months in quick-strip can during 16-month period (10) × 2 if increase in share or × 3 if decrease in share

[g] Line 9 + line 10

[h] Line 11 × line 12

Explanatory Note for D_1S_4

The assumption was made that Maxwell House share drops from 21.5 to 20.0 in Maxwell House quick-strip markets in 4 months after quick-strip can is introduced. Share remains at 20.0 for 2 more months, at which time Maxwell House reconverts to keyless can and share rises steadily from 20.0 to 21.5 in last 6 months

Δ Normal gross margin = Δ share points for last 10 months (−.9) × $116,667 × 10 months = $1,050,003

Δ Revenue due to Δ price = 20.38 × $24,419 × 4 = $1,990,637

Δ Can costs = $7,814 × 4 months × 20.38 = $636,997

Δ Promotion costs = extra amount M.H. would spend to regain lost share = incremental gross margin earned by price increase during period in which M.H. had quick-strip can ($1,991 − $637 = $1,354)

Explanatory Note for D_2S_2

Value of change in share beyond 16 months is 0. Assume that Maxwell House either would reconvert to keyless can or succeed in reducing costs to prevent a change in profit before taxes.

Explanatory Note for D_2S_3

The assumption was made that Maxwell House would switch back to keyless can after observing no change in its share of market for five months of quick-strip-can output. Simplifying assumptions were that reconversion is immediate and that it involves a cost small enough to be disregarded in this analysis.

Δ Can costs = $7,814 × 21.5 × 5 = $840,000

Explanatory Note for D_2S_4

The assumption was made that average Maxwell House share of market in markets in which quick-strip can is introduced drops from 21.5 to 20.9 4 months after the quick-strip can is introduced and stays at 20.9 for the next 2 months, at which time Maxwell House reconverts to the keyless can. Average share of market then steadily rises from 20.9 to 21.5 over the last six months of the 16-month period. Simplifying assumptions were that reconversion is immediate and that it involves a cost small enough to be disregarded in this analysis.

Average M.H. share of market for 4 months of quick-strip can output = 21.05

Average M.H. share of market for last 6 months (after reconversion to keyless can) = 21.2

Average M.H. share of market for last 12 months of 16 month period (starting with introduction of quick-strip can) = 21.1

Δ Normal gross margin = $116,667 × 12 months × .4 share points

Δ Can costs = $7,814 × 4 months × 21.05 = $657,939

Assume no additional promotion to regain lost share of market.

EXHIBIT 4–3

INPUT DATA SHEET FOR PRIOR ANALYSIS OF QUICK-STRIP-CAN DECISION BY COMPUTER WITH TIME-SHARING SYSTEM

You should fill out one input data sheet for each price alternative you wish to consider in your prior analysis. Enter the requested data in the spaces provided, taking care to observe decimal points.

Price change in dollars per pound [. _ _]

Number of Market-share profiles [_]

An expected outcome of the action of introducing the quick-strip can will be expressed in terms of a profile that represents Maxwell House's share of market over time. You can select three or four such profiles to represent the range of what you believe to be the possible outcomes of introducing the new can at a given price. The number of profiles you select is to be entered in the space provided above. You are about to be asked to supply several pieces of information about each of your market-share profiles.

MARKET SHARE AT POINT OF MAXIMUM CHANGE

Maxwell House's share of market at the time of decision (assume December 1, 1963) was 21.5. Introduction of the quick-strip can might lead to a change in the market share. In the spaces provided below, specify for each of your market-share profiles the value of the share at the point of greatest change from 21.5. This value can be any number from 0 to 99.9.

Profile 1	Profile 2	Profile 3	Profile 4
[_ _ . _]	[_ _ . _]	[_ _ . _]	[_ _ . _]

TIME OF REACHING POINT OF MAXIMUM CHANGE IN MARKET SHARE

Use a time code in which 0 stands for the beginning of output of Maxwell House ground coffee in the quick-strip can, 1.0 stands for 1 month later, and so forth. According to information given in Chapter 3, the total output of Maxwell House ground coffee could be offered in the new can for the first time at time 4.0, or 4 months after such output began. You now must specify for each market-share profile the point in time when you think the greatest change in Maxwell House's share of market under use of the quick-strip can would first be reached. For example, if you think this would be 6½ months after Maxwell House started to use the new container, you would enter 6.5. (Omit this input for a profile that represents no change from the 21.5 share of market.)

Profile 1	Profile 2	Profile 3	Profile 4
[_ _ . _]	[_ _ . _]	[_ _ . _]	[_ _ . _]

(In arriving at the next four inputs, assume that you would continue with the quick-strip can even if Maxwell House's share of market declined. This assumption will be relaxed later.)

NUMBER OF MONTHS MAXIMUM CHANGE IN MARKET SHARE IS HELD

You are to enter in the spaces below the number of months you think the maximum change in market share would be in effect. (Omit this input if profile represents no change from the 21.5 share of market. If you think maximum change in share would be held indefinitely, enter 99.9.)

EXHIBIT 4–3 (Continued)

Profile 1 Profile 2 Profile 3 Profile 4

POINT IN TIME WHEN INCREMENTAL GROSS MARGIN DUE TO PRICE INCREASE IS LOST

When Maxwell House offers the quick-strip can at a price increase, it may enjoy an increase in its gross margin. At some later time, however, sufficient competitive ground-coffee volume may be offered in the quick-strip can to prompt the use of special promotional deals and/or price reductions, so that the incremental gross margin which resulted from the price increase would be lost to Maxwell House Specify the point in time at which you think this would occur. (Use the same time code given earlier. Omit this input if price alternative represents no change. If you think incremental gross margin from price increase would be held indefinitely, enter 99.9.)

Profile 1 Profile 2 Profile 3 Profile 4

POINT IN TIME WHEN CHANGE IN MAXWELL HOUSE SHARE OF MARKET DUE TO QUICK-STRIP CAN BECOMES COMPLETELY DISSIPATED

At some point in time in the future, any change in share of market due to the quick-strip can will cease to have value to the Maxwell House management. Using the time code given earlier, indicate the point in time at which you think this will occur for each of your market-share profiles. (If you prefer to value a change in share indefinitely, enter 99.9.)

Profile 1 Profile 2 Profile 3 Profile 4

PROBABILITY OF MARKET-SHARE PROFILE OUTCOME

The market-share profiles you are using represent the range of possible outcomes of adopting the quick-strip can. In the spaces provided below, assign a probability of occurrence to each of the profiles you are using in the analysis.

Profile 1 Profile 2 Profile 3 Profile 4

WOULD YOU RECONVERT?

If the extent and duration of a change in share of market for a given profile would cause you to decide to reconvert to the keyless can, place a 1 in the box for the profile concerned. Otherwise, put a 0 in the appropriate box.

Profile 1 Profile 2 Profile 3 Profile 4

TIME OF CONVERSION

Using the time code given earlier, specify the time at which you would make the decision to reconvert to the keyless can. (Omit input if you would not reconvert.)

Profile 1 Profile 2 Profile 3 Profile 4

COST OF RECONVERSION

Enter in dollars your estimate of the total of additional costs that would be associated with the reconversion, including costs of any special promotions and

EXHIBIT 4-3 (Continued)

loss of plant efficiency. (Omit input if no reconversion.) For purposes of this exercise, assume that the cost of labor and materials for reconverting the plants would be $100,000. In thinking about additional costs of promotion, assume that Maxwell House spent about $4.5 million for advertising and $8.0 million for promotional deals in the previous fiscal year.

Profile 1	Profile 2	Profile 3	Profile 4
☐☐☐☐☐☐☐☐.☐	☐☐☐☐☐☐☐☐.☐	☐☐☐☐☐☐☐☐.☐	☐☐☐☐☐☐☐☐.☐

NUMBER OF MONTHS NEEDED TO RECAPTURE ORIGINAL SHARE OF MARKET

For each profile for which you would decide to reconvert, enter your estimate of the number of months that would pass between the time of the reconversion decision and the time at which the original 21.5 share of market would be recaptured or the time when additional costs associated with reconversion would cease. You are limited to a maximum input of 99.9 months. (Omit input if no reconversion.)

Profile 1	Profile 2	Profile 3	Profile 4
☐☐.☐	☐☐.☐	☐☐.☐	☐☐.☐

EXHIBIT 4-4

```
MAXWELL HOUSE COFFEE PRIOR ANALYSIS


PLEASE ENTER THE NUMBER OF PRICE ALTERNATIVES

1
PLEASE ENTER BELOW THE NUMBER OF PROFILES
YOU WISH TO USE IN ANALYSING EACH PRICE ALTERNATIVE

3
WHAT PRICE INCREASE IN DOLLARS ARE YOU  CONSIDERING

.02
ENTER THE PROBABILITIES FOR EACH PROFILE
PROBABILITY FOR PROFILE 1

.4
PROBABILITY FOR PROFILE 2

.3
PROBABILITY FOR PROFILE 3

.3
NOW FOR EACH PROFILE YOU WILL ENTER THE FOLLOWING
INFORMATION UNDER THE APPROPRIATE CODE
          A.  MAXIMUM MARKET SHARE
          B. CODE FOR TIME WHEN MAXIMUM CHANGE WILL OCCUR
          C. NUMBER OF MONTHS MAXIMUM IS HELD
          D. CODE FOR TIME GROSS MARGIN IS LOST
          E.  CODE FOR TIME WHEN EFFECTS ARE DISSIPATED
          F. INDICATION OF DESIRE TO RECONVERT("0" IF NO, "1" IF YES)
```

EXHIBIT 4-4 (Continued)

```
ENTER YOUR INPUTS FOR PROFILE  1
 A

24.
 B

6.5
 C

4.
 D

12.5
 E

18.5
 F

0
IS INPUT CORRECT ^ ( 0 IF NO, 1 IF YES )

1
ENTER YOUR INPUTS FOR PROFILE  2
 A

21.5
 D

14.5
 E

18.5
 F

0
IS INPUT CORRECT ^ ( 0 IF NO, 1 IF YES )

1
ENTER YOUR INPUTS FOR PROFILE  3
 A

20.
 B

8.5
 C

5.
 D

8.5
 E

99.99
 F

1
IS INPUT CORRECT ^ ( 0 IF NO, 1 IF YES )

1
SINCE YOU ARE RECONVERTING YOU MUST GIVE
          G. CODE FOR TIME YOU WILL RECONVERT
          H. COST IN DOLLARS TO  RECONVERT
          I. NUMBER OF MONTHS NECESSARY TO RECAPTURE ORIGINAL SHARE
```

EXHIBIT 4–4 (Concluded)

```
 G

13.5
 H

1250000.
 I

15.
IS INPUT CORRECT ^ ( 0 IF NO, 1 IF YES )


1
          PRICE ALTERNATIVE       1


         INCREASE IN PRICE IS    0.0200
```

PROFILE	DUE TO MKT SHARE	DUE TO PRICE INCREASE	CHG. IN CAN COST	LOST INCRE-MENT COSTS	RECON-VERSION COSTS	CHANGE IN PROFIT
1	3281259.	10398159.	-3327787.	-2235044.	-0.	8116587.
2	0.	9711461.	-3108018.	-1427770.	-0.	5175673.
3	-2931257.	6747967.	-2159594.	-1660197.	-1250000.	-1253081.

```
        EXPECTED VALUE OF ALL PROFILES IS    4423410.00
```

CHAPTER 5

EVALUATING RESEARCH PROPOSALS: PREPOSTERIOR ANALYSIS

The alternatives considered in Chapter 4 represent actions on the quick-strip can that Maxwell House executives could have taken on the basis of available information. The alternatives to be evaluated here call for seeking additional evidence on probable consumer demand before a final decision is made on whether to convert to the new can on a national basis. These alternatives are as follows:

A_{3a} Conduct a three-month sales test in Muncie and Stockton of the one-pound quick-strip can with a price increase of two cents per pound; then decide whether to proceed with adoption of the new can. Place order immediately for the equipment needed for plant conversion.

A_{3b} This is the same as A_{3a} except that results of the sales test would be reviewed as a basis for deciding whether to order the equipment needed for converting plant to quick-strip can.

A_{4a} Conduct three inquiries: the sales test planned for A_{3a}, consumer interviews in the test markets, and an in-home use test. Decide on basis of the research results whether to proceed with adoption of the new can. Place order immediately for the equipment needed for plant conversion.

A_{4b} This is the same as A_{4a} except that results of the inquiries would be reviewed as a basis for deciding whether to order the equipment needed for converting plant to quick-strip can.

Estimated costs of the research were $6,000 for the sales test in A_3 and $43,000 for the three inquiries in A_4. It is possible to reason that a decision should be made to conduct the proposed research, especially the sales test, because the cost is low; because an important decision made with research is more easily defended than one made without it; and because the research might turn up something of significance. Although it would be possible to go ahead for one or more of these reasons, that would beg the question of whether the research is likely to be worth more than it would cost, which research plan is the better bet, and whether Maxwell House should delay ordering the equipment needed for plant conversion until the research has been completed. Answers to these questions will be developed by preposterior analysis.

THE VALUE OF PERFECT INFORMATION

Before evaluating the research proposals, let us first determine the maximum amount of money the company would be warranted in spending for information that would permit choice of the best action with certainty. This amount is what is referred to as the value of perfect information. It can be ascertained by comparing the expected value under uncertainty of the alternative judged best by prior analysis with the expected value of that alternative under certainty. In the case of A_1, the value of perfect information would be equal to the cost of uncertainty involved in offering the quick-strip can at a price increase of two cents per pound. The figure is \$262,500 (.25 × \$1,050,000). (See Figure 4–3.)

Actually, of course, marketing research usually yields imperfect information, the value of which cannot exceed the cost of the uncertainty that it eliminates. The strategy for evaluating a research plan consists of (1) determining in advance of the research how the prior probabilities would be revised on the basis of the expected new information and (2) using the revised prior probabilities in calculations to ascertain whether uncertainty would be reduced enough to justify the research expenditure.

PREPOSTERIOR ANALYSIS OF A_{3a} WITHOUT THE COMPUTER

The proposed three-month sales test would use thirty-five retail stores that account for most of the coffee volume in Stockton and Muncie. Ten audits of ground-coffee sales would be made, one before the test began and the remainder at two-week intervals. Stockton represented the West, in which Maxwell House's market position was relatively weak. Muncie had a somewhat higher share of market for Maxwell House than was average for the United States. (The company had recent test-marketing experience in both cities.) Three months was the minimum time the marketing research manager believed to be needed for obtaining a meaningful indication of the effect of the test variables of the new can and the price increase. Regular output of Maxwell House coffee in the quick-strip can could not begin for four months at the earliest, so results of the sales test would be available in time to permit the company to abandon plans for changing to the new can if such action were indicated. The availability of only 150,000 one-pound cans before mid-March limited any test scheduled before then.

Conduct of the sales test could be combined with immediately ordering the equipment needed for plant conversion to the quick-strip can (A_{3a}) or delaying the equipment order until the test results were in (A_{3b}). Speculating with an early equipment order would run the risk of a loss if the sales test indicated that the new can should be rejected. The risk initially seems attractive, however, because the subjective probabilities used in the prior analysis indicate a high probability of success and a high opportunity cost of delay. Therefore, A_{3a} will be evaluated and the outcome reviewed to determine whether A_{3b} should be analyzed in detail.

The research focused on a price increase of two cents per pound, the alternative of no change in price having been eliminated by the prior analysis described in Chapter 4.

A DECISION TREE BRANCH FOR A_3. A tree structure suitable for evaluating either A_{3a} or A_{3b} is shown in Figure 5–1. The analysis involves using the sales-test findings to revise the prior probabilities for the quick-strip can. Alternative ways of making the revision will be discussed in this chapter. The first approach that will be described calls for answers to the following questions:

1. What observations of change in Maxwell House's share of market might be made in the proposed test?

2. What is the probability of making each of the above observations, conditional on the realization of each of the outcomes expected for the quick-strip can in the national market?

How well these questions can be answered depends on the extent of relevant past experience, on how well such past experience has been recorded and interpreted, and on the quality of judgment exercised in relating that experience to the specific proposal being considered. In this case, company executives had considerable test-marketing experience and recently had witnessed the change from the old key can to the keyless can for ground coffee. Considering the nature of the proposed change, there was reason to expect substantial homogeneity among markets in consumer reaction to it. Although empirical bases for answering the above questions typically are limited, such judgments are implicit in any evaluation of a market-test proposal. The Bayesian approach described here provides for making them explicit and incorporating them into the analysis.

The decision diagram in Figure 5–1 provides for considering what action should be taken assuming, in turn, that each of four representative results were observed in the sales test. The results are to be expressed as percentage changes from pretest share-of-market bench marks. To keep the example simple, Muncie and Stockton findings are to be weighted equally, combined, and treated as a single test. Actually, the two sets of findings should be inspected as separate entities before being aggregated

FIGURE 5–1

DECISION TREE BRANCH FOR A_3

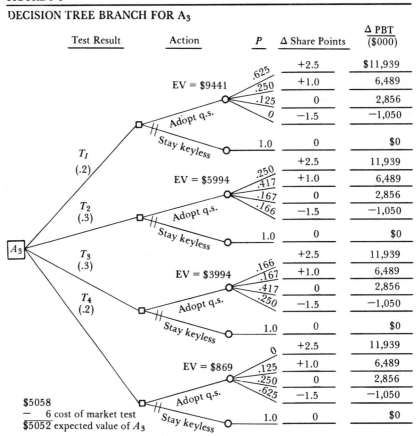

Test Result	Action	P	Δ Share Points	Δ PBT ($000)
T_1 (.2)	Adopt q.s. (EV = $9441)	.625	+2.5	$11,939
		.250	+1.0	6,489
		.125	0	2,856
		0	−1.5	−1,050
	Stay keyless	1.0	0	$0
T_2 (.3)	Adopt q.s. (EV = $5994)	.250	+2.5	11,939
		.417	+1.0	6,489
		.167	0	2,856
		.166	−1.5	−1,050
	Stay keyless	1.0	0	$0
T_3 (.3)	Adopt q.s. (EV = $3994)	.166	+2.5	11,939
		.167	+1.0	6,489
		.417	0	2,856
		.250	−1.5	−1,050
	Stay keyless	1.0	0	$0
T_4 (.2)	Adopt q.s. (EV = $869)	0	+2.5	11,939
		.125	+1.0	6,489
		.250	0	2,856
		.625	−1.5	−1,050
	Stay keyless	1.0	0	$0

$5058
− 6 cost of market test
$5052 expected value of A_3

in order to prevent possible loss of information; and if they were found to be considerably different, it could be argued that they should be formally analyzed separately.

The tree diagram assumes that a terminal decision of adoption or rejection will be made upon completion of the test. Although Maxwell

House executives might elect to undertake additional research at that time, a course of action providing for further inquiries, conditional on certain market-test observations, would constitute another alternative for evaluation.

Steps to be taken in the preposterior analysis will be discussed in the sections that follow.

SPECIFY POSSIBLE SALES-TEST OBSERVATIONS. For purposes of illustration, we shall assume that the following changes in Maxwell House's share of market represent the range of possible observations, considering the conditions of the test. Each value is a midpoint of a segment of the total range of values and is an average for the period of the test.

$$T_1 = +15\%$$
$$T_2 = +\ 7\%$$
$$T_3 = \pm\ 0\%$$
$$T_4 = -10\%$$

ASSIGN PROBABILITIES TO SALES-TEST OBSERVATIONS. A probability must be assigned for observing each test-market result, conditional on each of the expected outcomes of adopting the quick-strip can being realized in the nation as an average for the first ten months of quick-strip-can output.

The conditional probability assignments appear in Table 5–1. The table reflects the judgments, for example, that it O_1 (+2.5 share points) were to be the outcome of offering the quick-strip can in the nation, the probability of observing T_1 (+15 percent) in the market test is .5; the probability of observing T_2 (+7 percent) is .3; and so forth. Similarly, if O_3 (no change in market share) were to be realized in the nation, the probability of observing T_1 (+15 percent) in the market test is .1; the probability of observing T_2 (+7 percent) is .2; and so forth.

Table 5–1

CONDITIONAL PROBABILITIES OF POSSIBLE RESULTS OF MARKET TEST [a]

Outcomes for Nation: Change in Maxwell House Share Points	Market-Test Results: Change in Maxwell House Share of Market			
	T_1 (+15%)	T_2 (+7%)	T_3 (0%)	T_4 (−10%)
O_1 (+2.5)	.5	.3	.2	0
O_2 (+1.0)	.2	.5	.2	.1
O_3 (0)	.1	.2	.5	.2
O_4 (−1.5)	0	.2	.3	.5

[a] The values for changes in share of market are midpoints of ranges and are averages for the ten months of quick-strip-can output that Maxwell House would realize during the initial sixteen-month period of analysis.

The conditional probabilities are net expressions of a number of judgments about the accuracy and projectability of the test observations as indicated by the following questions:

1. What is the probability that the observations accurately represent what happened in the test market? Sampling and measurement errors and, sometimes, dishonesty of field audit personnel are among reasons why the probability is not 1.0.

2. What is the probability that what happened in the Muncie-Stockton test would have happened in the United States as a whole in the same time period? Are Muncie-Stockton ground-coffee buyers representative of coffee buyers generally with respect to their reactions to coffee cans and prices?

3. What is the probability that the full effect of the test variables of the new can and price was observed in the three-month period? Novelty appeal could produce larger sales in early months than could be sustained. On the other hand, more time might be required for measurement of the new can's ability to increase the likelihood of repeat purchase after continued use.

4. What is the probability that the test result would be indicative of what would happen with the two-pound quick-strip can? Only the one-pound size was available for the test.

These questions serve to point up the importance of accumulating empirical data to guide the valuing of research findings that is involved in making the conditional probability assignments. They also serve as reminders of the limitations of sales tests. In this case, for example, it would be possible to conclude that the results should be regarded only as a gross indicator of what might happen over a longer period of time if the quick-strip can were offered nationwide in both the one-pound and the two-pound sizes.

COMPUTE JOINT PROBABILITIES. The probabilities that each outcome for the nation and each market-test result would both occur appear in Table 5–2. They were computed by multiplying the prior probabilities for the national outcomes of offering the quick-strip can by the conditional probabilities of making the market-test observations. The formula is as follows:

$$P(O_i \cap T_j) = P(O_i)\, P(T_j|O_i) \qquad (i, j = 1, 2, 3, \ldots)$$

The probability that both O_1 and T_1 are true, for example, is $.25 \times .5 = .125$

REVISE THE PRIOR PROBABILITIES. The next step is that of computing the probability for each expected national outcome for the quick-strip can, conditional on observing each market-test result. The formula is as follows:

$$P(O_i|T_j) = \frac{P(O_i \cap T_j)}{P(T_j)} \qquad \text{(assuming } P(T_j) > 0\text{)}$$

If T_1 (+15 percent) were observed in the market test, for example, the revised probability for a national outcome of +2.5 share points from adopting the quick-strip can would be .125 divided by .2 = .625. Similarly,

$$P(O_2|T_1) = \frac{.05}{.20} = .25, \ldots$$

For illustrative purposes, the revised probabilities of outcomes of adopting the quick-strip can have been entered in the decision diagram (see Figure 5–1).

COMPUTE PROBABILITIES FOR TEST OBSERVATIONS. The probability for each of the research outcomes is obtained by adding the column of its joint probabilities in Table 5–2. The resulting sums have been entered in the decision diagram.

Table 5–2

JOINT PROBABILITIES OF OUTCOMES FOR NATION AND MARKET-TEST RESULTS [a]

Outcomes for Nation: Change in Maxwell House Share Points	Market-Test Results: Change in Maxwell House Share of Market T_1 (+15%)	T_2 (+7%)	T_3 (0%)	T_4 (−10%)	$P(O_i)$
O_1 (+2.5)	.125	.075	.050	0	.25
O_2 (+1.0)	.050	.125	.050	.025	.25
O_3 (0)	.025	.050	.125	.050	.25
O_4 (−1.5)	0	.050	.075	.125	.25
$P(T_j)$	.200	.300	.300	.200	1.00

[a] The values for changes in share of market are midpoints of ranges and are averages for the ten months of quick-strip-can output that Maxwell House would realize during the initial sixteen-month period of analysis.

COMPUTE CONDITIONAL EXPECTED VALUES OF ACTIONS. The expected value of a decision to adopt or reject the quick-strip can is computed by assuming, in turn, that each of the four market-test results was observed. In our example, delay is avoided by immediately ordering equipment needed for plant conversion. The sales test would be completed before regular output in the new can could begin.

With no delay involved, the expected changes in profits associated with the expected changes in share of market used in the prior analysis of A_1 with a price increase (see Figure 4–3) can be used here. They appear in Figure 5–1. Information on cost of converting the plant was not readily available; therefore, it was omitted from the computations. Under the assumptions, the cost of new equipment would be incurred in any event and, therefore, would affect all outcomes alike if it were included. Labor cost would not be incurred if conversion plans were abandoned, but such action was not indicated in the analysis.

If T_1 were observed in the test, the expected value of adopting the quick-strip can would be \$9,441,000. It was computed by multiplying each of the PBT outcomes of adoption by its revised probability and summing the products: (\$11,939,000 × .625) + (\$6,489,000 × .25) + (\$2,856,000 × .125) + (−\$1,050,000 × 0). Similarly, the expected value of the more favorable action was computed, assuming that each of the other test results was observed. The expected values appear in the tree diagram in Figure 5–1. It can be seen that even if the least-favorable observation, T_4, were made, the result was not valued heavily enough to lead to a rejection of the quick-strip can.

An assumption of no change in profit before taxes was sufficient for evaluating the action of remaining in the keyless can because that alternative could be rejected under the assumption. Actually, remaining in the keyless can could have a negative value if the quick-strip can helped competitors gain volume, partly at the expense of Maxwell House.

COMPUTE EXPECTED CHANGE IN PROFITS. This is done by (1) multiplying the four expected values calculated in the preceding step by the probability of their related test results, (2) summing the products, and (3) subtracting the cost of the test. In the example, the expected test observations would not remove uncertainty; therefore, the cost of the research could not be justified on the basis of the expected-value criterion. If the research plan had involved delay in converting to the quick-strip can, the expected changes in profits would have been less than those used in the prior analysis because at least some of the expected lead time over competition would have been lost. The expected value of the research alternative, therefore, would have been lower. Because this is apparent, a detailed analysis of A_{3b} is not needed, as it might have been needed if the prior probabilities had been less optimistic.

A few technical comments are helpful at this point. The expected value of the research alternative, A_{3a}, was less than that for the best nonresearch action, A_{1a}, by the amount of the direct costs of the research (see Figures 4–3 and 5–1). This kind of result will always be obtained under the following conditions: (1) if the research would not incur opportunity cost by delaying the terminal decision and (2) if no research observation would lead to a terminal action different from that which would have been taken in the absence of research. The second condition combined with opportunity cost, of course, would lead to a lower expected value.

In order for a research plan to have a higher expected value than the best nonresearch action, (1) one or more research observations must indicate a decision different from the one that would have been made on the basis of the prior analysis, and (2) the cost of uncertainty present in the prior analysis must be reduced by an amount exceeding the direct and the opportunity costs of research. One more note: if one or more research observations point to an action different from that indicated by the prior analysis, the expected value of the research plan will vary with changes in the conditional probability assignments for the research observations that would prompt a change in terminal action. A change in the mix of conditional probabilities (but not the total) assigned to research observations that would not lead to a change in terminal action would not affect the expected value of the research strategy.

ALTERNATIVE PROCEDURES FOR ARRIVING AT CONDITIONAL AND JOINT PROBABILITIES

The approach just described required the following judgments:

1a. Judgments of the range of changes in Maxwell House's share of the national ground-coffee market that might result from adopting the quick-strip can: O_i $(i = 1, 2, 3, \ldots)$
b. Judgments of the probability of occurrence of the anticipated changes in market share: $P(O_i)$

2a. Judgments of the range of changes in Maxwell House's share of market that might be observed in the sales test: T_j $(j = 1, 2, 3, \ldots)$
b. Judgments of the probability of observing each of the anticipated results in the sales test assuming, in turn, that each of the expected share-of-market outcomes would be realized in the nation over the remainder of the time period: $P(T_j|O_i)$

As shown earlier, these judgments were used to compute the joint probabilities of national and sales-test outcomes, $P(O_i \cap T_j)$, and by use of Bayes' theorem, to compute probabilities of the national outcomes conditional upon observing the expected text results

$$P(O_i|T_j) = \frac{P(O_i \cap T_j)}{P(T_j)}$$

Two important features of this procedure are that it (1) emphasizes reasoning from expected national outcomes to expected test-market observations and (2) employs judgments concerning net effect without requiring more specific thinking about the influence of each of the main factors contributing to the net effect. The question arises, therefore, of whether better estimates might be produced by modifying the approach with respect to one or both of the features just mentioned. Two alternative procedures will be described briefly.

MODIFICATION 1. This modification emphasizes reasoning from research observations to national outcomes. It consists of making the following judgments:

1a. Judgments concerning the changes in Maxwell House's share of market that might be observed in the sales test of the quick-strip can: T_j

b. Judgments concerning the probability of making each of the anticipated observations for the test: $P(T_j)$

2a. Judgments concerning the range of changes in Maxwell House's share of the national ground-coffee market that might result from adopting the quick-strip can: O_i

b. Judgments concerning the probability of realizing each of the national outcomes, conditional upon making each of the observations for the sales test: $P(O_i|T_j)$

The modified approach requires that probabilities be assigned to the anticipated test observations. In contrast, the approach used in the example on pages 71–77 required that probabilities be assigned to the possible national outcomes but not to the test observations. The latter probabilities emerged as sums of the joint probabilities for the test observations.

The choice between the procedures can be made on the basis of whether the decision maker, considering his experience and customary way of thinking, feels that he is better able to predict sales-test results or national outcomes. Choice of procedure may not make a great deal of difference when the two are expected to be very similar. It is more important in situations in which a substantial difference is expected.

MODIFICATION 2. This modification emphasizes reasoning from national outcomes to test observations, as in the approach used in the analysis of A_{3a}. It differs, however, in that it calls for reasoning separately about (1) the accuracy of the observed test-market results and (2) predicting a national outcome for a longer time period on the basis of true test results. The steps in the modification are as follows:

1. Estimate the range of true outcomes expected for the nation (the market area of concern): O_i.

2. Assign probabilities to the national outcomes: $P(O_i)$.

3. Estimate the range of true results expected for the market test: R_k.

4. Assign probabilities to each of the true results expected for the market test, conditional upon each of the anticipated national outcomes being true: $P(R_k|O_i)$.

5. Estimate the range of observations expected for the market test: T_j.

6. Assign probabilities for observing each of the anticipated results for the market test, conditional upon each of the results being true: $P(T_j|R_k)$.

7. Compute the joint probabilities for the different combinations of test-market observations, test-market true results, the national outcomes being true simultaneously: $P(T_j \cap R_k \cap O_i) = P(T_j|R_k)$ $P(R_k|O_i)$ $P(O_i)$. The equation assumes that $P(T_j|R_k) = P(T_j|$ R_k and $O_i)$.

8. Compute the joint probabilities of each of the market-test observations and each of the national outcomes both being true: $P(T_j \cap O_i) = \sum_k P(T_j \cap R_k \cap O_i)$.

9. Use Bayes's theorem to compute the probability for each of the national outcomes being true, conditional upon making each of the market-test observations:

$$P(O_i|T_j) = \frac{P(T_j \cap O_i)}{P(T_j)}$$

The values for $P(T_j)$ are produced by the joint probability computations in step 8.

Whether this approach should be used rather than the one described on pages 71–77 depends on whether it is likely to yield better probabilities of the anticipated national outcomes of the action being considered. Its potential advantage lies in separating out steps 4 and 6. Step 6 calls for judgments concerning the extent of sampling and measurement errors in the test-market observations. Estimating sampling error is often difficult in practice because of the necessity of using imperfect samples and the lack of complete knowledge of what the relevant variables are and how to allow for them. Step 4 calls for judgments about the representativeness of the markets and the time period used in the test.

Separating out steps 4 and 6 might encourage better recording and interpretation of experience so that the currently limited bases for making these judgments could be improved. In a good many companies, the marketing research department is best equipped to deal with steps 4 and 6, whereas the marketing (or brand or product) manager may be best qualified to estimate probable national outcomes of a given course of action.

Research is needed to learn more about which approach can give best results under different conditions and what division of labor is advisable in the making of the required judgments. Experiments could be run in which all three approaches were used. Theoretically, they should produce the same joint probability table: $P(O_i \cap T_j)$ in our example. In practice, it is very unlikely that the joint probabilities will be identical because of the impossibility of reasoning consistently about the uncertain net effects of a number of variables, some of which are unknown. Therefore, there can be value in getting judgments from more than one person and in using more than one questioning procedure on a given individual. Attempts to explain disparity of results should lead to the development of better procedures and should shake down executive thinking and improve the chances of making good decisions.

PREPOSTERIOR ANALYSIS OF A_3 USING THE COMPUTER WITH TIME-SHARING SYSTEM

A program for use with a computer time-sharing system was developed for preposterior analysis of the research alternatives. It is based on the model represented in Figure 5–1. Its input data sheet is given in Exhibit 5–1. A print-out of an analysis, including the inputs used as well as the results, is shown in Exhibit 5–2.

The preposterior analysis assumed that the research would involve no delay in moving toward adoption of the quick-strip can. Therefore, the market-share profiles and their probabilities, which were used in the prior analysis of A_{1a}, are applicable, and the changes in profits calculated for each profile in the prior analysis by computer were used as inputs in the preposterior analysis. Although the inputs were not intended to represent those used in the preposterior analysis made without the computer, a similarity is present. In both analyses, the least-favorable test result was not weighted heavily enough in the revision of prior probabilities to lead to rejection of the quick-strip can. Therefore, the value of the test-market alternative, A_{3a}, ($4,417,410) was the same as that for the action, A_{1a}, which would have been taken in the absence of the test ($4,423,410) except for the direct costs of the research ($6,000). The research alternative, then, would be rejected on the basis of the expected-value criterion being employed.

Availability of a time-sharing console makes use of the computer by executives feasible, convenient, and economical. The approach is straightforward, and several runs could easily be made using alternative conditional probability assignments to test the sensitivity of the outcome. The computational cost of the preposterior analysis run shown in Exhibit 5–2 was $2.10.

PREPOSTERIOR ANALYSIS OF A_{4a} WITHOUT THE COMPUTER

The high cost of delaying the final decision prompts consideration of ways of gaining as much information as possible in a short time. The alternative being considered here was designed with this objective in mind. It consists of the Stockton-Muncie test proposed in A_3, plus concurrent consumer interviews in the test market, plus an in-home use test of the quick-strip can.

The consumer interviews would be conducted by telephone among buyers of ground coffee in a random sample of telephone homes in Stockton and Muncie. Each respondent would be called four times at two-week intervals during the market test in order to obtain information on consumer awareness, attitudes, brand switching, and repeat purchasing in regard to the quick-strip can. About five thousand interviews would be made in the study, which would cost $12,000.

The in-home use test would take place in Detroit and San Francisco, markets in which Maxwell House had a high and a low share of the ground-coffee market, respectively. The respondents would be housewives who used ground coffee for at least half of their coffee requirements. Approximately half of them would be regular users of the Maxwell House brand. In the test, approximately seven hundred housewives would be asked to compare unbranded quick-strip and keyless cans containing coffee that would alternately be placed in the home for use at the beginning of each of eight weeks. At the end of the eighth week, the respondents would be asked which of the two cans they preferred and why, and what caused them to reject the can they favored least. They would be asked to rate each can on an overall basis and tell what they liked and disliked about it. They also would be asked which of the two cans they preferred on each of several qualities such as case of opening, ability to keep coffee fresh after opening, ease of closing, and safety. The test would cost $25,000.

The analysis of A_{3a} showed that the market test alone was not likely to be worth its cost. It remains to be seen whether the same is true for the combination of three inquiries that comprise A_{4a}.

The description of the procedure for the preposterior analysis of A_{3a}, including its possible modifications, is applicable here as well. Implementation in the case of a plan consisting of multiple inquiries that would produce many and varied items of information, however, entails special problems. Two of the studies in A_{4a} would produce verbal responses to questions put to consumers rather than share-of-market measures of buying behavior that would come from the market test. Relating such test results to national market outcomes of a given marketing action is difficult for reasons indicated earlier. Doing the same on the basis of data further removed from buying behavior is more difficult. The reason is that relatively little is known about what verbal responses to questions mean in terms of buying behavior.

Evaluation of the three-study plan of A_{4a} entails consideration of a number of questions such as the following:

1. How should each of the studies be evaluated? The market test would produce measures of percentage change in share of market. The four waves of interviews of respondents in the test markets and the interviews at the conclusion of the in-home use test would produce several kinds of responses such as statements of preference, awareness, past buying, buying intentions, and various spontaneous comments. Should all responses be equally regarded, or should weights be assigned to each kind of response? Or would it be better to review all evidence from a given study and then subjectively arrive at an evaluation of the total picture it represented?

2. Should the studies be evaluated as separate entities, or should their results be pooled with the evaluation being based on the aggregate? If there is to be pooling, should the data from the different studies be weighted? If so, on what basis?

3. Assuming that each study is to be evaluated as a separate entity, how should their evaluations be combined? The studies differ in several respects such as geographical location of respondents, size of sample, the kinds of questions asked, and the experience of respondents with the quick-strip can.

4. How should the evaluations of the findings of the three studies be incorporated into the analysis so that they can appropriately affect the prior probabilities for the expected national outcomes of offering the quick-strip can at a price increase of two cents per pound?

Typically, little valuable information is available from past experience when it comes to answering these questions. In the absence of sound empirical bases for making such judgments, the decision maker must do the best he can intuitively. Assumptions concerning the behavioral significance of different kinds of research findings must be made if research proposals are to be evaluated, regardless of what evaluative procedure is employed. The Bayesian approach has the value of being systematic in the identification and consideration of key questions and the existing bases for answering them.

A DECISION TREE FOR A_4. The decision diagram for A_4 in Figure 5–2 utilizes the same structure used for A_3. In both cases, possible research results must be identified. The prior probabilities then are revised assuming, in turn, that each research result were obtained. In the case of A_4, it is more difficult to arrive at net expressions of findings that have different meanings in terms of the direction and extent of the revision which should be made in the prior probabilities.

PROCEDURE FOR EVALUATION. The procedure described here involves one revision of the prior probabilities on the basis of the combined results of the three inquiries in the research plan. It includes the following steps:

1. For each of the three inquiries, categorize the possible outcomes in terms of whether they would cause you to leave your prior probabilities

FIGURE 5-2

DECISION TREE BRANCH FOR A_4

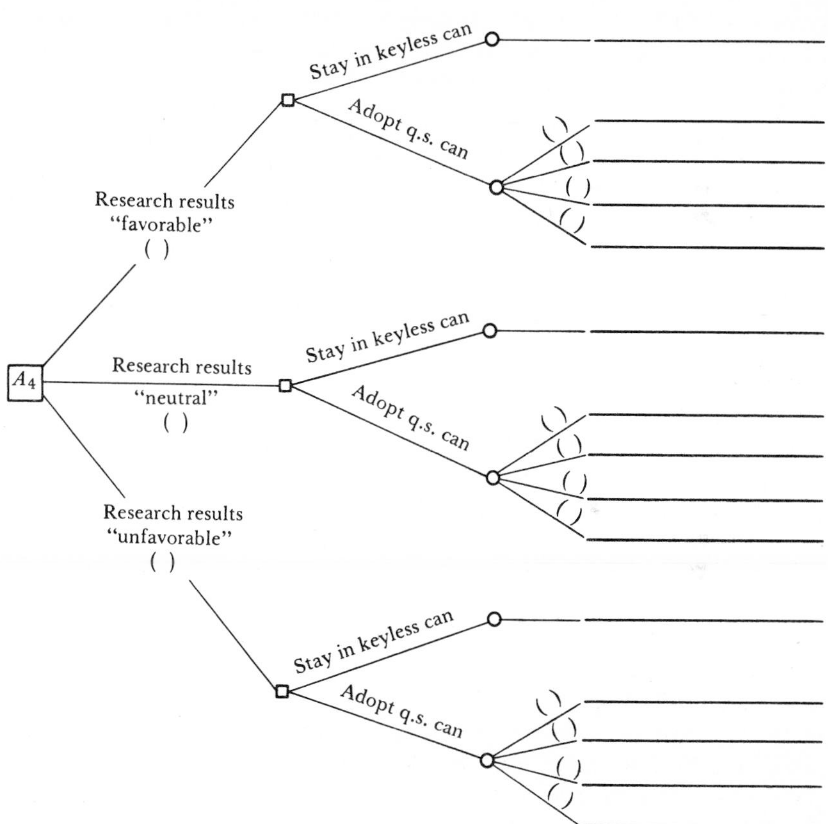

unchanged or to revise them either in a more optimistic or a more pessimistic direction concerning the outlook for the quick-strip can. The decision maker can use as many categories as he finds useful. The following three categories will be used in our illustration:

Category 1 Results that would prompt an optimistic revision of the prior probabilities for the quick-strip can

Category 2 Results that would leave the prior probabilities unchanged.

Category 3 Results that would prompt a pessimistic revision of the prior probabilities for the quick-strip can

2. Determine the relative weights that each of the three inquiries should have in the analysis. Note that in this example, they are weighted equally.

3. Group the possible combinations of the results of the three inquiries as being "favorable," "neutral," or "unfavorable," depending on whether they would prompt an optimistic revision of the prior probabilities for the quick-strip can, no revision, or a pessimistic revision. For example, the groups might be formed as follows:

	Number of Inquiries, by Category of Findings		
Group	Category 1	Category 2	Category 3
Favorable	3	0	0
	2	1	0
Neutral	2	0	1
	1	2	0
	1	1	1
	0	3	0
	0	2	1
Unfavorable	1	0	2
	0	1	2
	0	0	3

4. Assign the probability of observing each group of inquiry findings.

5. Relate the groups of inquiry findings to the anticipated outcomes (as identified in the prior probability distribution) of offering the quick-strip can. There are two ways of doing this (see pages 71–80). One method was used in evaluating the market-test alternative, A_3. It consisted of assuming, in turn, that each of the possible outcomes of adopting the quick-strip can would be realized and assigning a conditional probability to each of the anticipated test results, calculating joint probabilities, and computing revised probabilities of national outcomes conditional upon having observed the test results. An alternative approach is possible. It consists of reasoning from research results to national outcomes. The decision maker directly assigns the conditional probabilities of the national outcomes, assuming, in turn, that each of the groups of research results has been observed.

6. Using the revised probabilities for the national outcomes, compute the expected value for each alternative course of action (adopt or reject

quick-strip can), assuming, in turn, that each group of research results has been observed.

7. Compute the expected value for A_{4a} by (*a*) multiplying each of the three expected values obtained in the preceding step by the probability for its respective research-result group, (*b*) summing the products, and (*c*) subtracting the direct costs of the research.

An alternative procedure of three successive revisions of the prior probabilities, one for the results of each of the three inquiries, could be followed if there were reasons to believe it would lead to a better result. Usually, such reasons are not present because of marked limitations of bases for making the required judgments. If these limitations did not exist, theoretically the same revised prior probabilities should be obtained whether the decision maker relates national outcomes to research results as in the market-test example or relates research results to national outcomes. Usually, however, the bases are far from excellent. The decision maker should use the method that he believes is most likely to give the better result, considering the nature of his experience, his usual way of thinking, and the availability of information relevant to making the required judgments.

CATEGORIZING THE ANTICIPATED RESULTS OF THE MARKET TEST. To illustrate the approach (rather than to advocate any particular definitions), anticipated results of the Stockton-Muncie market test are categorized and defined as follows:

Category	Change in Share in Test Market If Test at $.02/lb Price Increase
Favorable	
Muncie	>+10%, or 1.1 share points
Stockton	>+15%, or 0.3 share points
Neutral	
Muncie	−5% to +10%
Stockton	−8% to +15%
Unfavorable	
Muncie	<−5%, or −.5 share points
Stockton	<−8%, or −.2 share points

Some of the judgments on which these definitions are based will be mentioned here to indicate relevant considerations. The required percentage changes in market share for Stockton are larger than those for Muncie because Maxwell House has a smaller position there; hence, a greater variability is to be expected in reported share-of-market figures. The definition of the neutral category assumes that prior probabilities should not be revised unless the market tests indicate a change from the current national market position of 21.5 share points of more than +10 percent or −5 percent. The judgment recognizes the limitations of the test mentioned earlier. Average share would be computed for each market from the six biweekly audits, omitting the first audit after the start of the test and any audit that appeared to be unusually out of line.

CATEGORIZING ANTICIPATED RESULTS OF TELEPHONE INTERVIEWS. Results of somewhat similar interviews conducted in the course of deciding to adopt the keyless can were summarized in Chapter 3. Although their value for judging the results of the proposed telephone interviews is questionable, they constitute the most applicable benchmark available. It is noteworthy that the earlier survey results were very favorable to the can being considered but that its later adoption was accompanied by very little if any increase in Maxwell House's share of market.

The following category-definition guidelines are offered in order to illustrate the approach rather than to advocate any particular definitions:

Category	Definition
Favorable	New can preferred over keyless can by 75 percent or more; noticeable switching to quick-strip can; no major dislikes about quick-strip (maximum of 15 percent negative response on any one item).
Neutral	From 50 to 75 percent prefer new can; few major dislikes about the quick-strip can; maximum of 30 percent negative response on any one item.
Unfavorable	Less than 50 percent prefer new can; noticeable switching away from quick-strip can; major dislikes about quick-strip can (at least 15 percent negative response on some item).

ASSIGN PROBABILITIES FOR GROUPS OF INQUIRY FINDINGS. A systematic approach to making the judgments that these probabilities represent can help maintain consistency of reasoning. One such approach consists of first establishing the probabilities that each of the three inquiries would produce category 1, category 2, and category 3 findings. These values can then be used to calculate the probabilities for observing the various groups of findings. Such an approach is illustrated in Exhibit 5–3, in which probabilities for the groups of findings were calculated on the assumption that any one of the inquiries had a .5 probability of producing category 1 findings, a .3 probability of producing category 2 findings, and a .2 probability of producing category 3 findings.

STEPS 5, 6, AND 7. These steps will not be executed here because they were illustrated in connection with the analysis of A_{3a}. Preposterior analyses made by the computer showed that it was very unlikely that A_{4a} would have a higher expected value than A_{1a}, given the prior probabilities used in our example.

PREPOSTERIOR ANALYSIS OF A_{4a} BY COMPUTER WITH TIME-SHARING SYSTEM

The program used in the preposterior analysis of A_{3a} can be used to evaluate any research plan. In the case of A_{4a}, the research outcomes would be the different groups (patterns) of the net expressions of the findings of the three inquiries instead of changes in share of market, as they were in the case of the A_{3a} market-test alternative.

STUDY QUESTIONS

1. On the basis of your own prior analysis of A_1 (see Chapter 3, Assignment 3), what is the value of perfect information?
2. Conduct a preposterior analysis of the action of following your research plan (see Chapter 4, Assignment 4), and then deciding whether to adopt the quick-strip can.
3. Do the results of your preposterior analysis agree with the conclusion indicated by the results reported in Chapter 5 that research should not be ordered? Why or why not?
4. Which procedure for revising prior probabilities do you prefer, the one followed in the example or the one in modification 1, as described in this chapter? Why?
5. How does the procedure for analyzing a research plan consisting of two or more inquiries (such as a market test and a consumer survey) differ from that for analyzing a plan consisting of a single inquiry?

EXHIBIT 5–1

INPUT DATA SHEET FOR PREPOSTERIOR ANALYSIS OF A_3 BY COMPUTER WITH TIME-SHARING SYSTEM

In your prior analysis of A_1 (Maxwell House decides to convert completely to the quick-strip can as soon as possible), you computed expected values for offering the quick-strip can at each of several alternative prices. Because of time limitations, you were not asked to analyze A_2 (Maxwell House stays with its present keyless can unless a major competitor switches to the quick-strip can. If this should happen, Maxwell House then would decide whether to adopt the quick-strip can).

In the preposterior analysis, you are to compare the following courses of action: (1) decide, on the basis of available information, to offer the quick-strip can as soon as possible at the price that gave the highest expected value in your prior analysis; (2) order research to obtain additional information for use in deciding whether to adopt the quick-strip can. In the process, you can analyze different research plans. You will need the inputs specified below.

EXHIBIT 5-1 (Continued)

Number of research plans to be considered ☐

Number of market-share profiles used in prior analysis ☐

EXPECTED CHANGE IN PROFIT AND PROBABILITY FOR EACH PROFILE

In the spaces below, enter for each profile the expected change in profit before taxes (PBT) in dollars and the probability of its occurrence.

Profile	Expected Change in PBT	Probability
1	☐☐☐☐☐☐☐☐☐.	.☐☐
2	☐☐☐☐☐☐☐☐☐.	.☐☐
3	☐☐☐☐☐☐☐☐☐.	.☐☐
4	☐☐☐☐☐☐☐☐☐.	.☐☐

If a proposed course of action that includes research involves following the same time schedule for contracting with the can company for quick-strip cans as that assumed in your prior analysis, the profiles, their probabilities, and their consequences in terms of expected change in profit before taxes will be the same as given in the prior analysis. If the proposed course of action that includes research specifies a different time schedule, new profiles and probabilities should be used in a new prior analysis to compute the expected changes in profit before taxes needed as inputs for the preposterior analysis.

You should prepare the following inputs for each research plan you wish to evaluate.

Direct costs of research in dollars ☐☐☐☐☐☐☐

Direct costs refers to all costs related to the research except for any opportunity cost of delaying a terminal decision until the research has been completed.

Number of possible research outcomes ☐

You are to select from two to four outcomes to represent the full range of possible observations that might be made in the proposed research. The number of research outcomes you wish to consider in the analysis should be entered in the space provided above.

CONDITIONAL PROBABILITIES FOR THE RESEARCH OUTCOMES

You now are to assign the conditional probability of observing each research outcome, assuming, in turn, that each of the possible market-share profiles you considered in your prior analysis would be realized if Maxwell House offered the quick-strip can at the price alternative specified for that market-share profile.

For example, you first should assume that market-share profile 1 would be realized if Maxwell House were to offer the quick-strip can at the price you specified for that market-share outcome. Under the assumption of market-share profile 1 being true, you are to assign the probability of observing R_1 in the proposed research. The probability should be entered in the appropriate space (top line of first column below). Then you should enter the probability of observing R_2, conditional on profile 1 being true (top line of the second

EXHIBIT 5–1 (Concluded)

column), and so forth. Repeat the process until you have assigned all probabilities required for the number of market-share profiles and research outcomes you are considering in the analysis.

Research Outcomes

Profile	R_1	R_2	R_3	R_4
1	[. | |]	[. | |]	[. | |]	[. | |]
2	[. | |]	[. | |]	[. | |]	[. | |]
3	[. | |]	[. | |]	[. | |]	[. | |]
4	[. | |]	[. | |]	[. | |]	[. | |]

EXHIBIT 5–2

```
MAXWELL HOUSE CASE-PREPOSTERIOR ANALYSIS

WE ARE NOW GOING TO CONSIDER WHETHER TO
SEEK ADDITIONAL INFORMATION FOR USE IN DECIDING
WHETHER TO ADOPT THE QUICK-STRIP CAN
HOW MANY RESEARCH PLANS ARE YOU CONSIDERING?
1
ENTER THE NUMBER OF PROFILES YOU USED IN PRIOR ANALYSIS
3
CHANGE IN PROFIT FOR PROFILE  1
8116587.0
PROBABILITY OF PROFILE  1
.4
CHANGE IN PROFIT FOR PROFILE  2
5175673.0
PROBABILITY OF PROFILE  2
.3
CHANGE IN PROFIT FOR PROFILE  3
-1253081.0
PROBABILITY OF PROFILE  3
.3
FOR RESEARCH PLAN  1  INDICATE IT S DIRECT COST IN  DOLLARS
6000.0
FOR RESEARCH PLAN  1 INDICATE THE NUMBER
OF POSSIBLE RESEARCH OUTCOMES YOU WANT TO CONSIDER IN ANALYSIS
3
NOW INDICATE BELOW YOUR ESTIMATE OF THE
PROBABILITY OF OCCURANCE FOR EACH POSSIBLE
RESEARCH OUTCOME, CONDITIONAL ON EACH MARKET
SHARE PROFILE BEING TRUE.
READ P(R=I/S=J) AS THE PROBABILITY OF RESEARCH I OCCURRING
GIVEN THAT MARKET PROFILE J IS THE TRUE STATE
```

EXHIBIT 5–2 (Continued)

```
P(R=1/S=1)
.7
P(R=2/S=1)
.2
P(R=3/S=1)
.1
P(R=1/S=2)
.2
P(R=2/S=2)
.6
P(R=3/S=2)
.2
P(R=1/S=3)
.0A
P(R=2/S=3)
.1
P(R=3/S=3)
.9
RESEARCH        PROFILE         REVISED PRIOR
OUTCOME                         PROBABILITY
 1                 1                0.824
 1                 2                0.176
 1                 3                0.0
THE EXPECTED VALUE OF ADOPTING THE CAN IF RESEARCH   1
IS OBSERVED IS       7597602.00


RESEARCH        PROFILE         REVISED PRIOR
OUTCOME                         PROBABILITY
 2                 1                0.276
 2                 2                0.621
 2                 3                0.103
THE EXPECTED VALUE OF ADOPTING THE CAN IF RESEARCH   2
IS OBSERVED IS       5321915.00


RESEARCH        PROFILE         REVISED PRIOR
OUTCOME                         PROBABILITY
 3                 1                0.108
 3                 2                0.162
 3                 3                0.730
THE EXPECTED VALUE OF ADOPTING THE CAN IF RESEARCH   3
IS OBSERVED IS        802355.62


EXPECTED VALUE OF RESEARCH PLAN  1 IS      4417409.00
```

EXHIBIT 5–3

CALCULATION OF PROBABILITIES OF OBTAINING EACH OF THE THREE GROUPS OF FINDINGS IN THE INQUIRIES OF A_4

Number of Inquiries, by Category of Findings 1	2	3	Multiplication [a]	Probability for Combination of Categories [b]
Favorable Group				
3	0	0	.5 × .5 × .5 × 1 =	.1250
2	1	0	.5 × .5 × .3 × 1 =	.2250
				.3500
Neutral Group				
2	0	1	.5 × .5 × .2 × 3 =	.1500
1	2	0	.5 × .3 × .3 × 3 =	.1350
1	1	1	.5 × .3 × .2 × 6 =	.1800
0	3	0	.3 × .3 × .3 × 1 =	.0270
0	2	1	.3 × .3 × .2 × 3 =	.0540
				.5460
Unfavorable Group				
1	0	2	.5 × .2 × .2 × 3 =	.0600
0	1	2	.3 × .2 × .2 × 3 =	.0360
0	0	3	.2 × .2 × .2 × 1 =	.0080
				.1040

[a] It was assumed that the probability of the findings of each inquiry falling in category 1 was .5; in category 2, .3; and in category 3, .2.

[b] The probability of obtaining a given combination of outcome categories from the three inquiries was obtained by multiplying the probabilities for the categorization together and then multiplying by the number of possible ways that combination of categories might be obtained. For example, there are three ways in which a combination of two category 1 findings and one category 3 finding might result.

Section Two

A Decision Problem for Prior Analysis

This section is concerned with decisions on product development and the positioning of new product and brand concepts in an existing line. Although it focuses on prior analysis alone, the material is more complex than that of the preceding section and requires more time to comprehend. Section Two places greater emphasis than Section One on the identification and screening of alternative courses of action, the structuring of a formal analysis, and the making of the managerial judgments required by the analytical model and the input data. More attention is also given to the use of the computer and the use of multiple decision criteria. The material is presented to allow the reader, if he so desires, to participate in the various steps by working on his own before he goes on to read in Chapter 7 about what someone else did.

CHAPTER 6

PLANNING THE DEVELOPMENT AND MARKETING OF A MAJOR PRODUCT INNOVATION: THE DECISION CONTEXT

A central objective of technical research efforts of the Maxwell House Division of the General Foods Corporation had been to find and develop means of improving the flavor characteristics of soluble (instant) coffee. A number of significant product improvements had been generated by these efforts, with the result that Maxwell House Division soluble-coffee brands had enjoyed a consistent quality advantage over competitors during the 1950s. At the same time, no practical means had been discovered for producing a soluble coffee that would match ground coffee in flavor.

During 1960, however, evidence accumulated indicating that a breakthrough had been achieved by an entirely new manufacturing process. Soluble coffee produced by the new process in a laboratory-scale plant was being rated as good as percolated ground coffee by expert tasters and by consumers in taste tests in which the coffees being compared were not identified.

The new coffee, given the code name V2, was made by a process based on the freeze-drying of concentrated brewed coffee to form soluble solids, as compared with the conventional spray-drying process then in use throughout the industry. Adapting the freeze-dry concept to soluble-coffee production had turned out to be a difficult task. The construction and operation of the laboratory-scale plant alone had represented a considerable technical achievement. Despite the favorable results of the laboratory plant, there was a good deal of uncertainty regarding the eventual technical and commercial feasibility of the process for a commercial-scale operation. On the other hand, the dramatic taste advantages of the V2 product and the possibility of establishing a lead time of perhaps three to five years over competition were persuasive arguments for immediate full-scale investigation of the feasibility of the process for commercial production.

In mid-1961, a decision was made to proceed directly to the construction of a commercial-scale plant to manufacture a V2 product. In January, 1962, with construction under way, management was formulating goals and priorities to facilitate the further technical development of the process and product and the development of marketing plans.

In this chapter, information is presented on (1) the United States coffee market, (2) the Maxwell House Division's position in the coffee market, and (3) the development of the V2 product. This information is intended to serve as a basis for outlining a reasonable set of alternatives for the further development and marketing of the V2 product(s).

THE UNITED STATES COFFEE MARKET

Total coffee volume in the United States in 1961 was estimated at 164,000,000 units[1] or about 1.32 units for every person over 14 years of age (see Exhibit 6–1). Total volume had increased an average of about 4 percent per annum in the previous eight years and per capita volume had increased at the slower average rate of about 2 percent per annum over the same time period.

Total coffee volume was split into two categories—ground and soluble—with the former accounting for about 113,000,000 units in 1961 and the latter accounting for 51,000,000 units. The relationship between the two was expressed by the so-called soluble ratio, which was the percent of total volume (in units) accounted for by soluble coffee. The soluble ratio had increased from 11 in 1953 to 31 in 1961 but had showed signs of leveling off in the more recent years (see Exhibit 6–1).

Maxwell House Division management thought that the major impediments to a higher rate of growth in per capita consumption of coffee were: (1) increasing competitive activity from other beverage categories, (2) concern with the possible adverse side effects of coffee on general health, and (3) the apparent nonacceptance of coffee as a warm weather, thirst-quenching beverage in the face of a general trend in society toward more emphasis on sports and outdoor living. They thought that the reasons for the declining rate of growth of soluble-coffee consumption would include (1) the fact that most consumers thought that

[1] A *unit* was an arbitrary physical measure of coffee volume. One unit was equal to twelve pounds of ground coffee or three pounds of soluble coffee. A unit of ground coffee so defined yielded approximately the same number of cups of coffee as a unit of soluble coffee so defined.

soluble coffee did not taste as good as ground coffee (for example, only 15 to 20 percent of those coffee drinkers who drank instant coffee exclusively thought that it actually tasted better than ground coffee), and (2) a prevalent opinion among consumers that there was "a certain social negative" in serving soluble coffee (for example, instant-coffee usage was thought to be a sign of a "lazy housewife").[2]

There were marked regional differences in the kinds of coffee consumed. In 1961, the soluble ratio in General Foods's eastern sales region was about 39 as opposed to a soluble ratio of 20 in the western sales region (see Exhibit 6–2). The factors that contributed to this difference had not been clearly identified.

In studies of coffee-consumption habits, coffee drinkers were typically classified into at least three groups: (1) ground users or those who used ground coffee more than 70 percent of the time, (2) soluble users or those who used soluble coffee more than 70 percent of the time, and (3) users of both who divided their consumption between ground and soluble coffee more evenly than the consumers in the first two categories did. On a national basis, ground users comprised 52 percent of all coffee drinkers; soluble users, 30 percent; and users of both, 18 percent. In Exhibit 6–3, coffee users are cross-classified according to the above groupings, by region, and by level of consumption. Further cross-classifications appear in Exhibit 6–4. Exhibit 6–5 indicates that about 75 percent of ground-coffee users in the United States had tried at least one brand of soluble coffee. Studies showed that women made about 70 percent of all brand selections of coffee.

Brand switching appeared to be more prevalent among soluble-coffee users than among ground-coffee users. A study conducted for the Maxwell House Division in San Diego in 1958 and 1959 indicated that 64 percent of the soluble-coffee users were hard-core users (that is, purchased 75 percent of coffee needs in one brand during a six-week period); whereas about 75 percent of ground-coffee users were hard-core users.

Coffee prices at retail had been moving down over the period from 1956 to 1961, from an approximate $0.98 per pound of ground coffee and $0.52 per two ounces of soluble coffee in 1956 to an approximate $0.70 per pound and $0.33 per two ounces for ground and soluble coffee, respectively, in 1961 (see Exhibit 6–6).

Chain and "super" independent grocery stores accounted for over 75 percent of retail coffee sales in the United States in 1961 (see Exhibit 6–7).

Data on competitors and competitive activities will be given in the

[2] For a published study on this point, see Mason Haire, "Projective Techniques in Marketing Research," *Journal of Marketing* (April, 1950): 649–656. Later studies conducted by General Foods tended to support Haire's results.

following sections. The reader who desires further general information on the coffee market and consumer motives and habits might refer to the Pan-American Coffee Bureau cases.[3]

POSITION OF THE MAXWELL HOUSE DIVISION IN THE COFFEE MARKET

In 1961, the Maxwell House Division sold three brands of coffee: Maxwell House, Yuban, and Sanka. Each brand had a ground and a soluble type; in addition, the Sanka brand was decaffeinated. The Division's share of total United States coffee sales was 35.1 percent in 1961, up from 23.9 percent in 1955. Maxwell House was the dominant brand, with the ground and soluble types holding, respectively, 14.0 percent and 14.4 percent of the total United States coffee market in 1961. Details on share of the national market held by the three brands over the years are given in Exhibit 6–8.

The position and trends in market share for the Maxwell House Division and its major competitors on a national basis are given in Exhibit 6–9 for the ground-coffee market and in Exhibit 6–10 for the soluble-coffee market. Figures on a national basis could be somewhat misleading, however, for there were sharp differences in the relative positions of the various brands and in competitive activity in different geographical areas. One of the most outstanding of these regional differences was between the East and West as illustrated by the following summary figures for 1961:

	East	West
Population (percent of total United States)	60%	40%
Total coffee consumption	95 million units	69 million units
Ground-coffee consumption	58 million units	55 million units
Soluble-coffee consumption	37 million units	14 million units
Soluble ratio	39	20
Maxwell House Brand Share of Ground- or Soluble-Coffee Market		
Ground Maxwell House	34.2%	8.2%
Instant Maxwell House	43.7%	34.5%
Maxwell House Advertising and Promotional Expenses Per Unit[4]		
Ground Maxwell House	$.30	$ .75
Instant Maxwell House	.80	1.60

[3] Joseph W. Newman, *Motivation Research and Marketing Management* (Boston: Division of Research, Graduate School of Business Administration, Harvard University, 1957), chapter VI, pp. 156–222.

[4] Promotional expenditures and gross-margin figures in the case are disguised.

Private brands offered by the various grocery chains were major competitive factors in sales through those channels. Of total sales through chain stores in 1961, the chain brands held about 30 percent of the ground market and about 50 percent of the soluble market; the corresponding figures for all General Foods's brands were 25 percent and 17 percent (see Exhibit 6–11).

In discussing the coffee business in general, Maxwell House management made the following observations:

1. Attractive gross profits (on the order of \$2.00 a unit for ground coffee and \$3.00 a unit for soluble coffee) and attractive returns on investment were being earned by coffee roasters and marketers.

2. There was significant overcapacity in coffee production.

3. Competition was extremely aggressive, particularly in the West, where roasters apparently obtained the bulk of their profits from ground coffee and appeared willing to operate their soluble business at relatively low profit.

4. The industry had developed to the point where, from the consumer's point of view, there was great similarity in product quality, packaging, and advertising and promotional techniques, particularly with respect to the soluble brands.

5. The number of brands of coffee being sold, particularly of the soluble type, was increasing each year. Further increases in competitive activity and new soluble-brand introductions were expected in the future. One facet of increased competitive activity was likely to be the expansion of brands with strong regional sales volumes into other regions of the United States. For example, periodic rumors of preliminary market investigations by Folger's had been received from Detroit, Milwaukee, and other Maxwell House sales districts where Folger's was not yet sold.

POSITION, OBJECTIVES, AND STRATEGY FOR INSTANT MAXWELL HOUSE

Instant Maxwell House was regarded as something of a phenomenon in the coffee industry. In the early 1950s, the brand had capitalized on its superior quality to capture a commanding position in the soluble-coffee market. This position had been maintained in later years despite narrowing quality differentials and aggressive advertising and promotional campaigns on the part of competitive manufacturer's brands and the lower-priced chain brands.

Excellent distribution had been achieved for Instant Maxwell House by 1961 with the two-, six-, and ten-ounce jar sizes selling, respectively, in 97, 96, and 87 percent of all United States grocery stores.

Advertising and promotion expenditures for Instant Maxwell House were, respectively, $0.35 and $0.73 per unit in 1961, for a total of $1.08 per unit. This figure was substantially below the estimated average of $1.37 per unit for all major brands in the soluble-coffee market. Details on advertising and promotion expenditures for Maxwell House Division soluble brands and all soluble brands are given in Exhibit 6–12.

Soluble-coffee marketers had tended to place increasing emphasis on promotions and less emphasis on media advertising. The majority of the promotional dollars were going to *off-label* deals.[5] The percentage of Instant Maxwell House sales made in off-label deals had increased from 15.8 percent in 1953 to 40.4 percent in 1961. This move by Maxwell House had been necessary to combat heavy dealing activity initiated by other brands. By 1961, sales on off-label promotions accounted for an estimated 90 percent of Folger's soluble sales and 70 percent of Nescafe's soluble sales (see Exhibit 6–13).

The broad marketing objectives established for Instant Maxwell House for 1962 were: (1) to develop the high volume and profit potential in the East and (2) to protect brand share from erosion in the West. The general strategies adopted to meet these objectives were: (1) in the East, maintain heavy advertising pressure to reinforce the strong position of the brand and utilize relatively light promotion rates in the hope of avoiding retaliatory increases in promotional activity on the part of competitors, and (2) in the West, maintain relatively light advertising pressure, but use heavy promotion to achieve shelf-price parity with leading competitors.

The general objective set for Instant Maxwell House advertising was to increase sales by attracting consumers to the brand (particularly from the ranks of ground-coffee users because of the higher per-unit gross profits in soluble coffee) while at the same time holding its position with present users of the brand.

The basic advertising strategy was to appeal to ground-coffee users on the theory that such advertising would also encourage users of competitive soluble brands to switch to Instant Maxwell House and reinforce the loyalty of current Instant Maxwell House users. Copy was based on the assumption that the most important characteristic desired by consumers in a cup of coffee was flavor and that the cup quality of ground coffee was the standard by which most coffee drinkers judged flavor. Specific copy objectives for Instant Maxwell House were: (1) to position Instant Maxwell House as a high-quality "real" coffee appropriate for all coffee-drinking occasions, (2) to inform consumers that despite superior quality, Instant Maxwell House sold at a popular price level,

[5] The term *off-label* referred to special tags or over-printing on product labels offering a special price reduction, for example, "10¢ off" or "20¢ off." In promotions of this type, the manufacturer absorbed the drop in price, and retailers received their regular dollar profit margin.

(3) to promise flavor superiority over competitive brands and support this promise with a believable reason, (4) if possible, to promise an advantage such as convenience over ground coffee, and (5) to set Instant Maxwell House advertising apart from all other coffee advertising. The emphasis on flavor appeals is evident in the following comments of the Instant Maxwell House product manager:

> We have told the Instant Maxwell House flavor story over the years with words descriptive of coffee properties such as "Fresh Flavor," "Freshest Taste in Coffee Yet," "Freshness and Flavor of Good Coffee," and "Rich Flavor and Aroma." . . . We have tried to employ copy claims expressive of both "real" coffee and flavorful coffee such as "Coffee Beans Fresh Hot from the Roaster." Competitors have tried to copy us, witness Nescafe's "43 Beans," Folger's "Bolder Beans," and Hills Brothers "Beans Selected by the Hills Family" campaigns. . . . The current problem is to search for new material that will continue to distinguish Instant Maxwell House advertising from that of its competitors. . . . At the present we are using the hyperbole "Cup and a half of Flavor" to attract consumers attention.

The objectives of Instant Maxwell House promotional efforts were: (1) to build volume by attracting a significant number of buyers who would not usually purchase the brand and (2) to protect market share by countering competitive activities. Maxwell House management reasoned that competitors could and would match price deals and, where possible, would match other kinds of promotional effort. Thus, price deals were used primarily as a defensive measure to prevent competitive retail prices from getting so far below Instant Maxwell House prices that previously loyal users would be lost and, less frequently, in an offensive move to provide extra value inducements to encourage switching to Instant Maxwell House. Aside from price dealing, major efforts were made to develop promotions that capitalized on Instant Maxwell House's market position and the technical capabilities of the General Foods organization. These promotions, such as those featuring reusable containers for packaging (carafes, thermoses) were fairly difficult for competitors to match directly.

POSITION, OBJECTIVES, AND STRATEGY FOR GROUND MAXWELL HOUSE

It was noted earlier that there were large regional differences in the brand share held by ground Maxwell House. Overall marketing objectives and strategies for 1962 were designed with these differences in mind.

In general, in those areas where Maxwell House had a large share of the ground-coffee market, the objectives for the brand were to protect this position and, where consistent with moderate marketing expenditures, to increase market share. The strategy to increase share relied primarily on special promotional campaigns in selected areas and an effort, through advertising and promotion, to increase sales of the two-pound pack of coffee.

In those areas where ground Maxwell House held a small percentage of the market, the objectives for the brand were to build market share in regions where competition was not exceptionally strong and to hold the existing position and remain competitive in those areas where vigorous regional competition was prevalent. The strategy for building market share was based on heavy promotional effort at both the trade and consumer levels and, on occasion, heavy regional advertising support to accompany the promotions. In those areas where the objective was to hold the brand position, the strategy was basically to meet competitive efforts but to go no further because in these areas the regional competitors were capable of and willing to meet and perhaps better any intensification of Maxwell House marketing effort.

POSITION, OBJECTIVES, AND STRATEGY FOR GROUND AND INSTANT YUBAN

Yuban Coffee had been developed and introduced as a premium coffee differing from other brands in flavor and price. Ground Yuban had a flavor judged by consumers to be richer and more full-bodied than other ground coffees; Instant Yuban's flavor was judged to be "bitter/burnt" and more like that of ground coffee than competitive soluble brands. Ground Yuban sold at about a 10-percent premium at retail over most manufacturer's brands, and Instant Yuban sold at about a 20-percent premium over most other soluble brands. Ground Yuban had been on the market for some years and in 1961 held 1.2 percent of the total coffee market, or about 1.8 percent of the ground-coffee market. Instant Yuban had been introduced in 1959 in selected areas, and expansion of its geographical coverage had been continuing since then. By 1961, Instant Yuban had achieved a 1.5-percent share of the total coffee market or 4.3 percent of the soluble-coffee market.

Research indicated that housewives who regularly purchased Yuban tended to be from households with annual family incomes of more than $7,000, to live in urban communities, to purchase both ground and soluble coffee, and to have considerable pride in their roles as housewives.

The marketing objectives for the Yuban brand in 1962 were: (1) to increase sales volume in the brand's top six districts, which were Boston.

Philadelphia, Los Angeles, New York, Syracuse, and San Francisco and (2) to maintain distribution and market share in other districts at about current levels. The strategy to achieve these goals was to concentrate advertising and promotion funds in the top six markets, with the emphasis to be placed on media advertising.

Yuban advertising was considered to be of central importance in building a premium image for the brand. The brand could not stand on its flavor qualities alone, for despite high-quality ingredients and agreement on its excellence by expert coffee tasters, many consumers could not discern any taste advantage for Yuban, particularly Instant Yuban, in blind taste tests. The advertising objectives for the brand then were to position it as *the* premium coffee and to convince all coffee drinkers that Yuban coffee was unmatched in flavor and quality. Copy strategy included: (1) emphasis on Yuban's higher price to reinforce the premium concept; (2) emphasis on Yuban's use of costly quality "aged beans" to support claims to flavor strength and superiority and to justify the higher price; and (3) capitalizing on Yuban product characteristics such as "bitter/burnt" flavor, higher caffein content, and "dark powder-roast," which consumer research had indicated were both distinctive of the brand and regarded favorably by coffee drinkers.

In the promotional area, both off-label deals and repetitive couponing were used periodically to attract and hold new users.

POSITION, OBJECTIVES, AND STRATEGY FOR GROUND AND INSTANT SANKA

Sanka was a decaffeinated coffee. Its market was defined as all current users of the brand plus those nonusers who had a moderate to high degree of concern about their caffein intake. Since Sanka's share of the total coffee market was about 4 percent, and since research indicated that about one coffee drinker in three was "concerned" about his caffein intake, there remained a considerable potential market for the Sanka brand. Two difficulties were paramount in increasing Sanka usage: (1) potential users tended to be distributed evenly by demographic characteristics and were thus difficult to single out for concentrated attention, and (2) although almost all prospects were aware of Sanka and its caffein-free benefits, they tended to believe that it was not real coffee and did not taste as good as real coffee. Research findings indicated that Sanka was used for the most part as a supplement to regular coffee consumption rather than as the primary coffee consumed in a household. In addition, there was survey evidence that nondecaffeinated-coffee users saw little difference between Instant Sanka and Decaf, the only significant competitive brand of decaffeinated coffee.

Emphasis in the Sanka marketing program was placed on convincing users and prospects that Sanka was 100 percent pure, good-tasting coffee with an added caffein-free benefit. It was presented as a coffee to be used and served like other coffees. Since Ground Sanka volume was small as compared with Instant Sanka, the major marketing effort was placed behind Instant Sanka. In 1961, Ground Sanka held 1.3 percent of the ground-coffee market, Instant Sanka held 8.8 percent of the soluble market, and Decaf held 1.6 percent of the soluble market.

Advertising and promotional expenditures for Instant Sanka in 1961 were $0.90 and $0.50 per unit, respectively. The Sanka product manager thought the low promotional rate was made possible by Sanka's unique attributes.

Sanka advertising strategy was to present Sanka as 100-percent pure coffee, to emphasize its fine flavor, and to assign to it all the sensory pleasures and emotional satisfactions to be gained from drinking coffee. At the same time, Sanka's distinct feature—that it was 97 percent caffein-free, allowing a person to drink as much as he liked—was woven into the advertisements. For example, one ad featured a large picture of a man drinking coffee and the copy: "Drink this hearty coffee as strong as you like. . . . it still can't get on your nerves." A more recent advertisement used a headline stating: "Indulge yourself. . . . get all the best of the coffee bean aroma, flavor, but not caffein!"

Sanka promotional objectives were to increase the rate of product trial among current nonusers and to increase the rate of purchase among those consumers who used Sanka intermittently. The emphasis in promotional activity was on couponing and other consumer-level promotional techniques aimed at stimulating product trial. Off-label deals were not used as frequently by Sanka as they were by other soluble-coffee brands.

AN OVERALL VIEW

Sizing up the situation, management had concluded that the Division was in need of a significant innovation to support additional growth and to protect its existing position. The reasoning was as follows:

1. Both the ground and soluble portions of the total coffee market were segmented as much as possible by existing Division entries; Maxwell House in the popular price field, Sanka in the decaffeinated field, and Yuban in the premium price field.

2. Further growth in share of market was not expected for the following reasons: (*a*) In the *ground market,* there probably was no acceptable blend or roast difference that could be used with existing brands to significantly increase their consumer preference. Existing regional brands had good products, long-established names, and were aggressive competitors. Promotional innovations such as a three-pound package size and reusable containers had been used by the Maxwell House Division

and matched by competition. (*b*) In the *soluble market,* the historical product-quality advantage enjoyed by Instant Maxwell House had been narrowed greatly and was dangerously close to parity with other brands as viewed by the consumer. Instant Maxwell House's large share of market was being attacked by more than four hundred competitive brands. Improved products and heavy promotional spending already had begun to cut into Instant Maxwell House's share in some regions and posed a severe obstacle to Instant Yuban's growth. In addition, the known marketing innovations for increasing soluble-coffee sales, such as aroma oils, giant-size packages, new jar shapes, and reusable carafe-packages, had been nearly exhausted.

The problem was summed up as that ". . . of having to reacquire a significant coffee product advantage in order to achieve an objective of increased profit." It was against this background that the decision was made to aggressively pursue the development of freeze-dried coffee.

DEVELOPMENT OF THE FREEZE-DRIED PRODUCT

The Maxwell House Division long had supported considerable technical research for soluble coffee. The research had paid off as evidenced by the quality advantage enjoyed by Instant Maxwell House over the years. At the present time, however, the technical people believed that the conventional spray-dry process for producing soluble coffee had been developed as far as was economically feasible.

In the spray-dry process, freshly roasted ground coffee beans were percolated in large tanks in much the same way as home-brewed percolated coffee was prepared. The brew was then concentrated by continued boiling and was sprayed into the top of large cylindrical towers. As the spray fell in the tower, it passed through a stream of hot air that evaporated the moisture and allowed the coffee solids to fall to the bottom. The basic problem insofar as coffee quality was concerned was the need to use fairly high temperatures that drove off many of the volatile elements responsible for coffee flavor. Attempts to recover the volatile elements and add them back to the coffee solids had met with some success, but there were technical and economic limits on the extent to which this modification could be pursued.

The freeze-dry process offered a way to avoid high temperatures and, in theory at least, a way of producing soluble coffee that would closely approximate ground-coffee flavor. In the freeze-dry process, concentrated percolated coffee was prepared in much the same way as in the spray-dry process. It was then flash frozen and passed into a vacuum chamber wherein the moisture content was removed by sublimation.[6] The result-

[6] *Sublimation* refers to the passage of a substance directly from its solid to a gaseous state.

ing product was small, dry crystals of percolated coffee that remained in the solid form indefinitely without refrigeration.

Research on the freeze-dry process had started in 1956. By 1961, a laboratory-scale pilot plant was producing about two pounds of the V2 product per day. Experimentation with the laboratory plant revealed that a broad spectrum of coffee tastes could be produced depending upon the ingredients and process conditions (for example, kind of beans used, amount of roasting, and yield) used in a production run. This was a considerable departure from experience with the spray-dry process, which had been dubbed "the great leveler" because of the relative insensitivity of its product to changes in ingredients and process conditions.

Taste experimentation had centered on twenty-seven different varieties of V2 products produced by twenty-seven different sets of ingredients and process conditions chosen to represent a reasonable sampling of the possibilities of the process. The following results for three V2 varieties illustrate the potential differences in product (the results are based on blind [7] taste tests among consumers):

	Percent Preferring Test Product Versus [a]	
Test Variety	Instant Maxwell House (Soluble Users)	Ground Maxwell House (Ground Users)
A	44 (N = 400)	58 (N = 340)
B	52 (N = 900)	50 (N = 750)
C	56 (N = 725)	41 (N = 520)

[a] These are disguised figures, but they roughly represent the situation.

The potential differences in product made possible by the new process and the apparent differences in taste preferences of soluble- and ground-coffee users raised important questions regarding the choice of formulation and market target (s) for further development work.

For a variety of reasons, it appeared that any V2 product would be considerably more expensive to manufacture than a spray-dried product. Estimates based on projections from laboratory experience were that early full-scale production costs would most likely be about $1.50 higher per unit than current Instant Maxwell House costs and that over time this differential might be reduced to about $1.00 per unit. These facts had spurred an effort to reduce costs by mixing the V2 product with the spray-dried product in the hope of retaining the flavor advantages of the V2 product while achieving economies through the use of the cheaper spray-dry process and the existing spray-dry facilities. Experimental

[7] In the blind taste test, the consumer had no information on the brands or types of coffee being tasted.

work had involved various mixes, and early taste tests had been encouraging. Although mixtures offered considerable economic potential, the technical problems of achieving a satisfactory product were extremely difficult.

For the purposes of this case, consideration of mixtures is limited to an assumed combination of 80-percent spray-dried and 20-percent freeze-dried coffee.

CONSTRUCTION OF A FULL-SCALE PLANT

In 1961, the issue of the next step in the development of the freeze-dried product had been raised. At that time, the research group thought that the safest move would be to scale up the laboratory plant to a production rate of about fifty pounds a day and to conduct further tests on the feasibility of the process and product. A number of difficult technical problems remained to be solved, and they reasoned that the problems would only be further complicated by jumping to a large-scale plant.

Maxwell House Division management concluded, however, that if the freeze-dried product were as good as it appeared to be, it would be better to move directly to the construction of a commercial-scale plant that would make possible a conclusive evaluation of the freeze-dry process and the marketing prospects of a freeze-dried product. In addition, the move would maintain or increase Maxwell House's lead time over competitors. In July, 1961, the General Foods Board of Directors approved an initial allocation of about $3,000,000 for the development and construction of a plant with a capacity of 100 units of freeze-dried soluble coffee per hour. The plant was expected to operate 120 hours per week for 52 weeks per year.

THE DECISION SITUATION

In January, 1962, work was proceeding on the design and construction of the commercial plant, which was expected to be in production by the end of the year. In the meantime, further research was being conducted using the laboratory plant.

For some time, Division executives had been visualizing and discussing a rather large number of alternative plans for the development and eventual marketing of the freeze-dried product or products. There were differences of opinion about the relative merits of some of the plans. The time had come, however, when decisions had to be made on specific objectives and priorities for guiding further work.

Chapter 6 concludes at this point, rather than becoming more specific about possible alternatives or about how the problem might be approached. This is done so that, at the end of each chapter, the reader may use what he has already learned about the decision problem to try to work through it on his own. It is this kind of experience that leads to an understanding of the operational aspects of applying decision theory in realistic business situations

STUDY QUESTIONS

ASSIGNMENT 1

1. What questions had to be answered in January, 1962, so that work on the development and eventual marketing of freeze-dried coffee could proceed without delay? In what order should the questions receive attention?

ASSIGNMENT 2

1. What alternative courses of action relative to the development and marketing of freeze-dried coffee might be followed? (*Course of action* refers to a sequence of steps to be taken within a specified period of time.)

2. How would you proceed to determine which of the possible alternatives should be subjected to detailed analysis?

3. What are the main uncertainties that must be considered in choosing among the alternative courses of action?

ASSIGNMENT 3

1. For purposes of this assignment, assume that freeze-dried coffee would be offered in only one formulation under one brand. Given this assumption, what alternative courses of action relative to the development and marketing of the brand were open to the Maxwell House Division?

2. Which of the alternative courses of action would you subject to formal analysis in the process of choosing among them? Why?

3. Draw a decision tree for use in analyzing the alternatives you selected for formal analysis.

ASSIGNMENT 4

1. How would you carry forward a formal prior analysis of the alternatives included in your decision tree?
   a. What measures would you use in comparing the alternatives?
   b. What kinds of calculations should be made?
   c. How would you arrive at the values of the various inputs needed for the calculations?

EXHIBIT 6–1

ESTIMATED UNITED STATES COFFEE VOLUME, 1953 TO 1961

Year[a]	Volume (in Million Units)[b]			Soluble Ratio[c]	Units Per Capita (over age 14)
	Total	Ground	Soluble		
1961	164	113	51	31	1.32
1960	156	109	47	30	1.33
1959	154	109	45	29	1.33
1958	148	107	41	28	1.28
1957	138	104	34	25	1.21
1956	132	104	28	21	1.20
1955	120	97	23	19	1.10
1954	123	106	17	14	1.15
1953	121	108	13	11	1.13

NOTE: The figures in all exhibits in Chapter 6 conform to the definitions given for this exhibit unless otherwise noted.

[a] General Foods's fiscal years, which begin in April.

[b] A *unit* was an arbitrary physical measure of coffee volume. One unit was equal to twelve pounds of ground coffee or three pounds of soluble coffee. A unit of ground coffee so defined yielded approximately the same number of cups of coffee as a unit of soluble coffee so defined.

[c] $\text{Soluble ratio} = \dfrac{\text{soluble volume in units} \times 100}{\text{total volume in units}}$

SOURCE: Adapted from Maxwell House Division records.

EXHIBIT 6–2

ESTIMATED REGIONAL COFFEE CONSUMPTION, 1955 TO 1961[a]
(In Million Units)

Year	Eastern United States Consumption			Western United States Consumption		
	Ground Coffee	Soluble Coffee	Soluble Ratio	Ground Coffee	Soluble Coffee	Soluble Ratio
1961	58	37	39	55	14	20
1960	55	34	38	54	13	19
1959	56	32	36	54	13	19
1958	54	29	35	52	12	19
1957	54	25	32	50	9	15
1956	55	21	28	48	7	15
1955	52	17	25	45	6	12

[a] The division into eastern and western regions was based on a classification of Maxwell House Division sales districts. The eastern districts included: Boston, New York, Philadelphia, Syracuse, Washington, D.C., Youngstown, Cincinnati, Louisville, Charlotte, Atlanta, Jacksonville, Memphis, New Orleans, Detroit, and Indianapolis. The western districts included: Chicago, Milwaukee, St. Louis, Minneapolis, Omaha, Kansas City, Dallas, Houston, Portland, San Francisco, Los Angeles, Denver, and Phoenix. Approximately 60 percent of the population was in the eastern region and 40 percent in the western region.

SOURCE: Adapted from Maxwell House Division records.

EXHIBIT 6–3

CLASSIFICATION OF COFFEE CONSUMERS, BY TYPE OF COFFEE CONSUMED, REGION, AND USAGE, 1960
(Percentages)

		Region		Level of Consumption		
Type of Coffee Consumed [a]	Total	East	West	Heavy (>5 cups per day)	Moderate (4, 5 cups per day)	Light (1, 2, 3 cups per day)
Ground users (>70% ground)	52	45	63	58	47	50
Users of both ground and soluble	18	19	17	13	20	21
Soluble users (>70% soluble)	30	36	20	29	33	29
All consumers	100	60	40	41	26	33

[a] "Ground users" used ground coffee more than 70 percent of the time; "soluble users" used soluble coffee more than 70 percent of the time; and "users of both" divided their consumption between ground and soluble coffee more evenly than consumers in the first two categories did. These definitions apply throughout Chapter 6 unless a notation appears to the contrary.

EXHIBIT 6-4

DEMOGRAPHIC CHARACTERISTICS OF CONSUMERS OF VARIOUS TYPES OF COFFEE, 1960
(Percentages)

	Categories of Coffee Consumers [a]				
Demographic Characteristic	Total Coffee Drinkers	Soluble Users	Instant Maxwell House Users	Users of Both Soluble and Ground	Ground Users
Age					
18–29	21	20	24	23	22
30–39	23	21	23	23	23
40–49	21	18	18	19	22
50 and over	35	41	35	35	33
Total	100	100	100	100	100
Socioeconomic					
Upper	8	8	10	12	8
Upper middle	15	12	19	21	15
Middle	43	44	40	40	41
Lower middle	17	17	13	14	18
Lower	17	19	18	13	18
Total	100	100	100	100	100
Amount Consumed					
Heavy	41	28	39	47	46
Moderate	26	26	28	29	23
Light	33	46	33	24	31
Total	100	100	100	100	100
Size of City					
Over 1,000,000	11	9	10	12	9
250,000–1,000,000	11	11	11	10	13
100,000–250,000	7	7	8	5	7
25,000–100,000	11	14	12	12	9
2,500–25,000	19	14	17	22	21
Under 2,500	10	11	9	8	11
Open country	31	34	33	31	30
Total	100	100	100	100	100

[a] See Exhibit 6–3 for category definitions.
SOURCE: Maxwell House Division records.

EXHIBIT 6–5

TRIAL OF SOLUBLE COFFEE BY GROUND-COFFEE DRINKERS, PRIOR TO 1961 [a]
(Percentages)

	Ground-Coffee Drinkers		
	United States	East	West
Number of Soluble-Coffee Brands Tried			
One brand	25	25	26
More than one brand	46	48	42
Unknown number of brands	2	2	2
Total	73 [a]	75	70
Brand of Soluble Coffee Tried			
Instant Maxwell House	53	61	38
Instant Folger's	14	1	39
Nescafe	19	21	15
Instant Sanka	19	19	18
Instant Chase & Sanborn	17	20	10
Hills Bros.	4	3	5

[a] Another study concluded that 92 percent of all exclusively ground-coffee drinkers had tried soluble coffee.
Source: Maxwell House Division records.

EXHIBIT 6–6

AVERAGE REGIONAL RETAIL PRICES FOR GROUND AND SOLUBLE COFFEE, 1956 TO 1961

	Eastern United States		Western United States	
Date	Ground Coffee (1 lb)	Soluble Coffee (2 oz)	Ground Coffee (1 lb)	Soluble Coffee (2 oz)
August 1, 1961	.70	.33	.69	.32
August 1, 1960	.72	.35	.73	.35
August 1, 1959	.72	.36	.74	.37
August 1, 1958	.87	.43	.87	.43
August 1, 1957	.96	.48	.97	.49
August 1, 1956	.98	.52	1.00	.54

Source: Maxwell House Division Records.

EXHIBIT 6–7

COFFEE SALES, BY TYPE OF GROCERY STORE, 1961

Type of Store	Percent of Total Number of Stores	Percent of Total Coffee Sales
Small independent	57	5
Medium independent	18	5
Large independent	13	13
Super independent	5	26
Chain stores	7	51

SOURCE: Adapted from Maxwell House Division records.

EXHIBIT 6–8

PERCENTAGE OF THE UNITED STATES TOTAL (SOLUBLE AND GROUND) COFFEE MARKET, BY GENERAL FOODS BRANDS OF COFFEE, 1955 TO 1962

Year	Ground Maxwell House	Instant Maxwell House	Ground Sanka	Instant Sanka	Ground Yuban	Instant Yuban	Total
1961	14.0	14.4	0.9	3.1	1.2	1.5	35.1
1960	13.7	14.4	0.8	3.1	0.7	0.3	33.0
1959	12.8	13.2	0.8	2.9	0.7		30.4
1958	11.8	12.7	0.7	2.6	0.7		28.5
1957	11.9	11.9	0.7	2.3	0.7		27.5
1956	13.0	10.7	0.6	1.8	0.6		26.7
1955	12.6	8.9	0.6	1.3	0.5		23.9

SOURCE: Maxwell House Division Records.

EXHIBIT 6-9

PERCENTAGES OF THE UNITED STATES GROUND-COFFEE MARKET, BY SELECTED BRANDS OF GROUND COFFEE, 1953 TO 1961

Year	Maxwell House	Sanka	Yuban	Folger's	Hills Bros.	Chase & Sanborn	Chains' Own Brands
1961	21.6	1.3	1.8	15.0	8.8	5.8	14.6
1960	20.8	1.2	1.0	14.8	8.6	6.0	14.9
1959	19.1	1.1	0.8	14.0	8.7	5.6	15.9
1958	17.3	1.1	0.8	13.0	9.0	5.1	16.3
1957	16.7	1.0	0.7	12.3	8.3	4.5	17.1
1956	17.3	0.9	0.5	13.0	8.7	5.2	16.1
1955	16.3	0.8	0.4	11.2	9.5	5.6	14.6
1954	16.5	1.0	0.4	11.4	NA [a]	NA	NA
1953	15.6	1.1	0.3	11.5	NA	NA	NA
Percentage distribution in grocery stores, 1961	94	94	27	54	44	74	45

[a] NA stands for "not available."
SOURCE: Maxwell House Division Records.

EXHIBIT 6–10

PERCENTAGES OF THE UNITED STATES SOLUBLE-COFFEE MARKET, BY SELECTED BRANDS, 1953 TO 1961

Year	Instant Maxwell House	Instant Sanka	Instant Yuban	All General Foods Brands	Nescafe	Chase & Sanborn	Chains' Own Brands	Instant Folger's	Decaf	All Others
1961	41.2	8.8	4.3	54.3	8.9	5.9	8.7	5.7	1.6	14.9
1960	42.2	9.0	0.8	52.0	9.6	6.0	9.1	4.4	1.7	17.2
1959	40.3	8.9		49.2	10.5	6.7	9.3	3.7	2.2	18.4
1958	40.8	8.4		49.2	11.6	8.5	8.2	3.3	2.4	16.8
1957	42.6	8.2		50.8	13.6	9.0	7.3	2.0	1.7	15.6
1956	43.8	7.3		51.1	16.9	5.8	4.8	2.5	1.2	17.7
1955	40.2	6.0		46.0	18.5	8.5	1.7	2.8	0.8	21.5
1954	34.6	6.5		41.4	21.8	11.9	0.3	1.5	0.3	23.1
1953	32.9	6.7		39.6	27.2	10.7				22.5

SOURCE: Maxwell House Division Records.

EXHIBIT 6–11

PERCENT OF CHAIN STORE COFFEE SALES, BY CHAIN STORE BRANDS AND GENERAL FOODS BRANDS, 1956 TO 1961

	Ground Coffee		Soluble Coffee	
Year	Chain Brands	General Foods Brands	Chain Brands	General Foods Brands
1961	30.1	24.6	50.1	16.5
1960	31.4	22.8	47.0	17.6
1959	34.2	20.2	44.3	18.0
1958	36.2	17.5	44.0	15.9
1957	NA[a]	16.5	45.0	22.4
1956	NA	16.0	46.8	18.6

[a] NA stands for "not available."
SOURCE: Maxwell House Division Records.

EXHIBIT 6–12

ESTIMATED PROMOTIONAL AND ADVERTISING SPENDING FOR MAXWELL HOUSE DIVISION SOLUBLE BRANDS AND ALL SOLUBLE BRANDS, 1958 TO 1961 [a]

Brand	1958		1959		1960		1961	
	Estimated Spending ($000,000) [b]	Unit Rate	Estimated Spending ($000,000)	Unit Rate	Estimated Spending ($000,000)	Unit Rate	Estimated Spending ($000,000)	Unit Rate
Instant Maxwell House								
Promotions	6.7	.40	7.8	.43	10.5	.53	15.3	.73
Advertising	6.8	.41	7.2	.40	7.9	.40	7.4	.35
Total	13.5	.81	15.0	.83	18.4	.93	22.7	1.08
Instant Sanka								
Promotions	1.0	.29	1.4	.34	1.3	.30	2.2	.50
Advertising	3.8	1.12	3.5	.87	4.3	1.02	4.1	.90
Total	4.8	1.41	4.9	1.21	5.6	1.32	6.3	1.40
Instant Yuban								
Promotions			.3	.45	3.1	.72	3.5	1.46
Advertising			.3	.48	1.9	.65	6.4	2.68
Total			.6	.93	5.0	1.37	9.9	4.14
All Solubles								
Promotions	18.9	.53	20.7	.57	25.3	.65	35.6	.82
Advertising	25.2	.72	24.3	.63	23.2	.58	23.5	.55
Total	44.1	1.25	45.0	1.20	48.5	1.23	59.1	1.37

[a] Figures disguised.
[b] ($000,000) stands for "in millions of dollars."
SOURCE: Maxwell House Division Records.

EXHIBIT 6–13

ESTIMATED PERCENTAGES OF ANNUAL SALES IN OFF-LABEL PROMOTIONS FOR SELECTED BRANDS OF SOLUBLE COFFEE, 1953 TO 1961

Year	Instant Maxwell House	Chase & Sanborn	Nescafe	Folger's	All Soluble Brands
1961	40.4	84.8	69.2	90.2	48.0
1960	23.7	82.5	59.9	71.3	35.0
1959	30.7	92.2	48.5	59.6	37.2
1958	33.6	76.6	27.0	37.3	33.3
1957	10.5	78.4	20.8	2.8	20.4
1956	8.2	37.3	23.5		16.3
1955	11.0	49.3	14.7		17.2
1954	0.0	55.2	12.0		
1953	15.8	13.6	12.3		

SOURCE: Maxwell House Division Records.

CHAPTER 7

IDENTIFYING AND SCREENING ALTERNATIVES AND STRUCTURING THE ANALYSIS

As pointed out in Chapter 1, Bayesian decision theory is most directly applicable to two of the five steps outlined for the decision-making process: (1) the evaluation of alternatives and (2) choice among alternatives. The problem of choice claimed most of our attention in the earlier chapters and will continue to do so. With its demand for specificity, the Bayesian approach indirectly can encourage systematic work on the equally important functions of identifying, formulating, and screening alternative courses of action for evaluation. Because of their importance to management, these steps will be discussed in this chapter before beginning the structuring of the analysis.

Looking beyond the present chapter to the work that lies ahead, more detailed information relative to the planning and the conduct of the analysis will be given in Chapter 8. Later chapters will deal with the development of the analytical model, the preparation of the input data it requires, the conduct of the prior analysis by computer, and the evaluation of the output.

KEY DECISIONS TO BE MADE

In view of the complexity of the freeze-dried-coffee problem, it is important to identify the inherent management decision questions before becoming too involved with details. Once this has been done, the nature of courses of action will become more apparent and the groundwork will have been laid for creating a framework for handling the analysis.

With construction about to start on a commercial-scale plant, management must decide what kind of freeze-dried product should be developed first. The answer depends, in turn, on the answer to a larger question: What program for the development and marketing of the innovation would be most profitable over future years? In other words, it is necessary at the outset to look ahead to determine, at least in broad outline, how freeze-dried coffee should be positioned in the Division's coffee line, which already includes three brands of both ground and spray-dried coffee.

A number of interrelated issues must be resolved in formulating a future program. One concerns the number of brand concepts that should be developed for freeze-dried coffee. There are several possibilities. For example, freeze-dried coffee might be introduced as (1) a new brand or brands; (2) a premium grade, supplementing the present grade, under one or more of the Division's existing brands; or (3) a replacement for the formulation now employed by one or more of the present brands. If two or more brands are to be offered, their sequencing as well as their timing must be determined. Perhaps one plan should be followed for the entire United States. Consideration of variations for the East and the West, however, is prompted by the marked geographical differences in market position of the Maxwell House Division regular and instant coffees.

A basic issue is whether the freeze-dried product(s) should be designed to appeal primarily to users of ground coffee, users of soluble coffee, or users of both. Results of taste tests cited in Chapter 6 showed that the different user groups reacted differently to test formulations.

A question with important technical and financial aspects is whether the company should try to develop and market a 100-percent freeze-dried product or a mix of 20-percent freeze-dried and 80-percent spray-dried coffees. Other product questions to be answered have to do with characteristics such as taste, color, and size of particles.

ALTERNATIVE COURSES OF ACTION

The number of theoretically possible alternative courses of action is huge. As suggested above, the alternatives would represent different mixes of such variables as the identities of brand concepts for freeze-dried coffee, the number of such concepts to be developed, and the timing and sequencing of the developmental work and marketing for each brand concept chosen. Additional alternatives might consist of different plans for different geographical regions.

To further emphasize the complexity of the problem, a partial list of alternatives is presented below. It is long even though it is arbitrarily limited by the assumptions that no more than two brand concepts for freeze-dried coffee would be developed and that they would be offered nationwide. Under these constraints, the following alternatives are possible:

1. Introduce one new brand
   a. A regular (with caffein) brand
   b. A decaffeinated brand

2. Add a premium grade under an existing brand
   a. A Super Instant Maxwell House
   b. A Super Instant Yuban
   c. A Super Instant Sanka

3. Replace the formulation of one existing brand
   a. A new Instant Maxwell House
   b. A new Instant Yuban
   c. A new Instant Sanka

4. Introduce two new brands
   a. A regular brand and a decaffeinated brand
   b. Two regular brands
   c. Two decaffeinated brands

5. Introduce one new brand, and add a premium grade under an existing brand
   a. A regular brand and a Super Instant Maxwell House
   b. A regular brand and a Super Instant Yuban
   c. A regular brand and a Super Instant Sanka
   d. A decaffeinated brand and a Super Instant Maxwell House
   e. A decaffeinated brand and a Super Instant Yuban
   f. A decaffeinated brand and a Super Instant Sanka

6. Introduce one new brand, and replace the formulation of one existing brand
   a. A new regular brand and a new Instant Maxwell House
   b. A new regular brand and a new Instant Yuban
   c. A new regular brand and a new Instant Sanka
   d. A new decaffeinated brand and a new Instant Maxwell House
   e. A new decaffeinated brand and a new Instant Yuban
   f. A new decaffeinated brand and a new Instant Sanka

7. Add a premium grade under two existing brands
   a. A Super Instant Maxwell House and a Super Instant Yuban
   b. A Super Instant Maxwell House and a Super Instant Sanka
   c. A Super Instant Yuban and a Super Instant Sanka

8. Add a premium grade for one existing brand, and replace the formulation of one existing brand

a. A Super Instant Maxwell House and a new Instant Yuban
b. A Super Instant Maxwell House and a new Instant Sanka
c. A Super Instant Yuban and a new Instant Maxwell House
d. A Super Instant Yuban and a new Instant Sanka
e. A Super Instant Sanka and a new Instant Maxwell House
f. A Super Instant Sanka and a new Instant Yuban

9. If this list were expanded to include alternatives that provided for undertaking the development of two concepts sequentially rather than concurrently, the number of alternatives under actions 4, 5, 6, 7, and 8 would be tripled. (There would be three possibilities for each of the combinations of two brand concepts: (*a*) concurrent development, (*b*) *A* before *B*, and (*c*) *B* before *A*.

It should be apparent that many more alternatives could be listed, each one representing a somewhat different pattern of development over a period of years. For example, the pattern might provide for the addition of more new brands, the later introduction of one or more premium grades under existing brands, or the later replacement of the formulations for existing brands. The list would become several times longer if the brand-positioning possibilities given above were combined with the alternative actions of offering a 100-percent freeze-dried product or a mix of freeze-dried and spray-dried coffee, and offering formulations that would appeal to different degrees to users of ground coffee, users of soluble coffee, and users of both ground and soluble coffee.

SELECTING ALTERNATIVES FOR DETAILED ANALYSIS

The preceding section suggests the need for an orderly means of determining which of the many alternatives deserve serious-enough consideration to be subjected to formal analysis. It is important at the outset to recognize the principal categories of possible actions so that promising ones will not be overlooked. It is neither practical nor desirable, however, to analyze all possibilities in detail.

Bayesian decision theory itself has nothing directly to say about how alternatives should be either identified or preliminarily screened. Instead, it specifies how the formal analysis should be conducted, given the alternatives from which a choice is to be made. The decision maker, using his experience and judgment, should reduce the list of possible alternatives as much as he feels he can. He probably will do better if he considers procedure before he plunges into the task. There is no one best way of proceeding. An approach that can be useful, however, consists of developing a set of questions which are critical to deciding the scale of developmental activity and choosing among proposed plans.

For purposes of illustration, questions that could be useful in the freeze-dried-coffee situation are listed below. Some of them relate to known (or ascertainable) practical constraints; whereas others relate to the uncertainties inherent in the decision situation.

PRACTICAL CONSTRAINTS. The following questions focus on possible resource limitations:

> How many projects could be handled in the next year by the present technical staff and laboratory facilities? How many later on?
>
> Should current staff and facilities be expanded now?
>
> What output limitations would be imposed by production facilities?
>
> Are there any financial constraints that would limit developmental activity?

UNCERTAINTIES. A number of uncertainties could influence the scale and pace of developmental activity as well as the choice of a specific course of action. They are reflected in the following questions:

> What are the chances that development can be successfully completed for a 100-percent freeze-dried product? A 20-percent freeze-dried and 80-percent spray-dried mix? How long would it take?
>
> What are the chances that a new freeze-dried product can be made successfully in a commercial-scale plant (so that it would compare as favorably with ground coffee in consumer taste tests as the freeze-dried product which was made in the laboratory did)?
>
> What demand would there be for a 100-percent freeze-dried product? A 20–80 mix product?
>
> > What segments of the market represent the best potential customers?
> >
> > How successful is the company likely to be in convincing people that freeze-dried coffee has superior taste qualities and that they should try it?
> >
> > How many people would be willing to pay a substantially higher price for the new product?
>
> How would the introduction of freeze-dried soluble coffee affect the sales of each of the existing Maxwell House Division brands?
>
> What are the chances that production costs for freeze-dried cof-

fee can be reduced enough to permit the new product to be sold profitably?

How much risk does the company wish to assume in developing freeze-dried coffee? What programs of development and marketing meet the risk criteria?

Our main interest here is in procedure, so no attempt will be made now to answer these questions in detail. Sufficient information was presented in Chapter 6, however, to permit the reader to formulate similar questions, answer some of them, and form useful judgments relative to others. All things considered, practical constraints and the uncertainties either force or argue for activity on some limited scale in the early stages.

STRUCTURING THE PROBLEM UNDER A SIMPLIFYING ASSUMPTION

In order to facilitate further handling of the problem, we shall temporarily impose the simplifying assumption that only one brand concept for freeze-dried coffee is to be developed.

ALTERNATIVES. A reasonably comprehensive set of alternative developmental programs under the assumption consists of various combinations of the following actions:

1. Aim development toward a product that will have:
   a. A strong appeal to users of ground coffee
   b. A strong appeal to users of instant coffee
   c. An appeal to both ground- and instant-coffee users

2. Use ingredients consisting of:
   a. 100-percent freeze-dried coffee
   b. A mix of 20-percent freeze-dried and 80-percent spray-dried coffee

3. Offer the new product as one of the following brands:
   a. A new regular (with caffein) brand
   b. A new decaffeinated brand
   c. An added Super Instant Maxwell House
   d. An added Super Instant Yuban
   e. An added Super Instant Sanka
   f. Instant Maxwell House
   g. Instant Yuban
   h. Instant Sanka

SCREENING THE ALTERNATIVES. Selection of alternatives for formal analysis involves considering how each one might affect the overall profits of the Maxwell House Division, recognizing the context of the market positions of the Division's present brands. Observations and speculations that might be made in the course of preliminary assessment including the following:

▷ *Market Target.* The Maxwell House Division has more to gain by winning over to its brands present users of ground coffee rather than users of soluble coffee. This is true, especially in the West, because of the Division's dominant position in the soluble-coffee market. At the same time, however, soluble-coffee users may be more inclined than ground-coffee users to try a new soluble coffee. A difficult question relative to defining the target market is whether a freeze-dried product designed to have some appeal to both groups (present ground- and soluble-coffee users) would do as well as or better than one designed to appeal more strongly to one group than to the other.

▷ *100-percent Freeze-Dried versus 20–80 Mix.* A 100-percent freeze-dried product has the better chance of product and market success and probably can be developed in a shorter period of time. The 20–80 mix, however, offers substantially lower costs of both capital equipment and production, and it could be sold at a lower price.

▷ *Regular versus Decaffeinated Product.* It is not clear which product would enjoy the larger sales volume. The regular soluble market is larger in total size; but the decaffeinated market is sizable, and there are few decaffeinated brands. There is reason to believe that the taste of spray-dried decaffeinated soluble coffee constitutes a formidable barrier to sales. A new brand that combined the advantages of decaffeinated and instant coffee with the taste of ground coffee might be able to gain a substantial sales volume faster than any other freeze-dried-coffee alternative. Many of these sales, however, probably would come at the expense of Sanka (both Instant and Ground). Development of a decaffeinated product would involve special technical problems.

▷ *New Brand, Regular Soluble Coffee.* A number of arguments can be advanced for this alternative. A new brand with a distinctive name, packaging, and advertising could represent the most effective means of communicating the fact of a major soluble-coffee innovation. It might be easier to get acceptance of a higher price with a new brand concept than with an existing brand. A new brand probably would involve less cannibalism of the sales of existing Maxwell House Division brands. A successful additional brand could give the Division the advantage of more shelf space in retail stores. The reputations of existing brands would not be jeopardized in the event that the new offering proved to be either a product or a marketing failure. A new brand, however, can be expected to have higher introductory marketing costs.

▷ *Replacement for an Existing Brand Formulation.* Use of an established brand name might lead to greater and more rapid market acceptance. An argument could be advanced that freeze-dried coffee eventually will be used in all soluble brands and that Maxwell House would enhance its position in the long run by adopting freeze-dried coffee for its existing brands as soon as possible. By doing so, it would enjoy maximum lead time over competitors for important product differentiation. Arguments against using freeze-dried coffee for the existing brands include the difficulty that might be encountered in going to higher prices and the probability indicated by taste-test findings that a number of people prefer the taste of the present formulations to that of the freeze-dried product.

▷ *A Premium Brand Supplement.* Using freeze-dried coffee as a premium brand supplement would make use of an established name to facilitate gaining acceptance at lower marketing costs. Cannibalism of the sales of the brand it supplemented, however, might be substantial. A premium supplement could be viewed as a transitional step toward discontinuing the present formulation in favor of offering only the freeze-dried product under the brand name at the higher price.

The above comments were not offered to advance any particular position. Instead, the purpose was to indicate the character of the early thinking that must be done by the decision maker. As he proceeds, he must become more specific. In screening each alternative course of action, he should consider the following factors, although no more than order-of-magnitude estimates of the values are necessary at this stage.

1. Probable sales volume

2. Probable gross margin

3. Probable marketing costs for introductory and ongoing operations

4. Cannibalism costs (that is, the effect on the sales and profit contributions of present Maxwell House Division brands)

5. Investment costs

6. Time estimates for each alternative covering product development, test marketing, and market introduction

7. Probability of developmental and marketing success

8. Probable timing and nature of competitive entry

BUILDING A DECISION TREE. One way of laying out the problem will be described here. It represents an attempt to picture the principal decisions, presenting them in the order of their basic importance (see Figure 7-1).

The two main branches of the decision tree represent alternative decisions about target markets. (The number was limited in the interest of simplicity.) One branch represents directing efforts primarily to the winning of customers from the ranks of present users of soluble coffee. The other defines the target market primarily in terms of the present users of ground coffee. Taste-test results reported in Chapter 6 indicated that somewhat different product specifications would be required for the two groups. This, in turn, could mean differences in the time necessary for successful development.

The next set of branches distinguishes between the types of product that might be employed regardless of the brand concept chosen. One is a 100-percent freeze-dried coffee, the other a mix of 20-percent freeze-dried coffee and 80-percent spray-dried coffee. The alternative brand concepts (A–E) chosen for formal analysis would appear as the branches on the right of the tree. They are considered individually because they would meet with different levels of consumer demand and have somewhat different implications for the product development program. For example, a product designed to replace the present formulation for Instant Maxwell House (which has been a popular price–mass market item) would have to meet tighter cost limitations than a product intended for marketing under a new premium brand would. If the formulation for an existing brand were to be replaced by a freeze-dried product, the latter would have to meet rigid specifications to avoid alienating present users of the brand.

Alternatives involving decaffeinated products would be integrated into the two broad market-target branches in this particular tree. A case can be made, however, for treating potential customers for a freeze-dried decaffeinated coffee as a third target market. The product would have unique developmental demands.

Because of technical difficulties yet to be overcome, the 20–80 mix was judged to have a much lower probability of developmental success than the 100-percent freeze-dried product. This difference is taken into account in the tree. It is handled by assuming certainty of developmental success for the 100-percent freeze-dried product and providing for an assignment of an appropriately lesser probability for the 20–80 mix. Although the probability of developmental success for the 100-percent freeze-dried product actually was less than certainty, it was regarded as being high enough so that the certainty assumption could be used satisfactorily in the relative analysis for choosing the better of the two mix alternatives.

If efforts to develop the 20–80 mix were to fail, a decision then could be made to attempt development of a 100-percent freeze-dried product.

FIGURE 7-1

DECISION TREE: ASSUMING DEVELOPMENT OF ONE BRAND CONCEPT FOR FREEZE-DRIED COFFEE

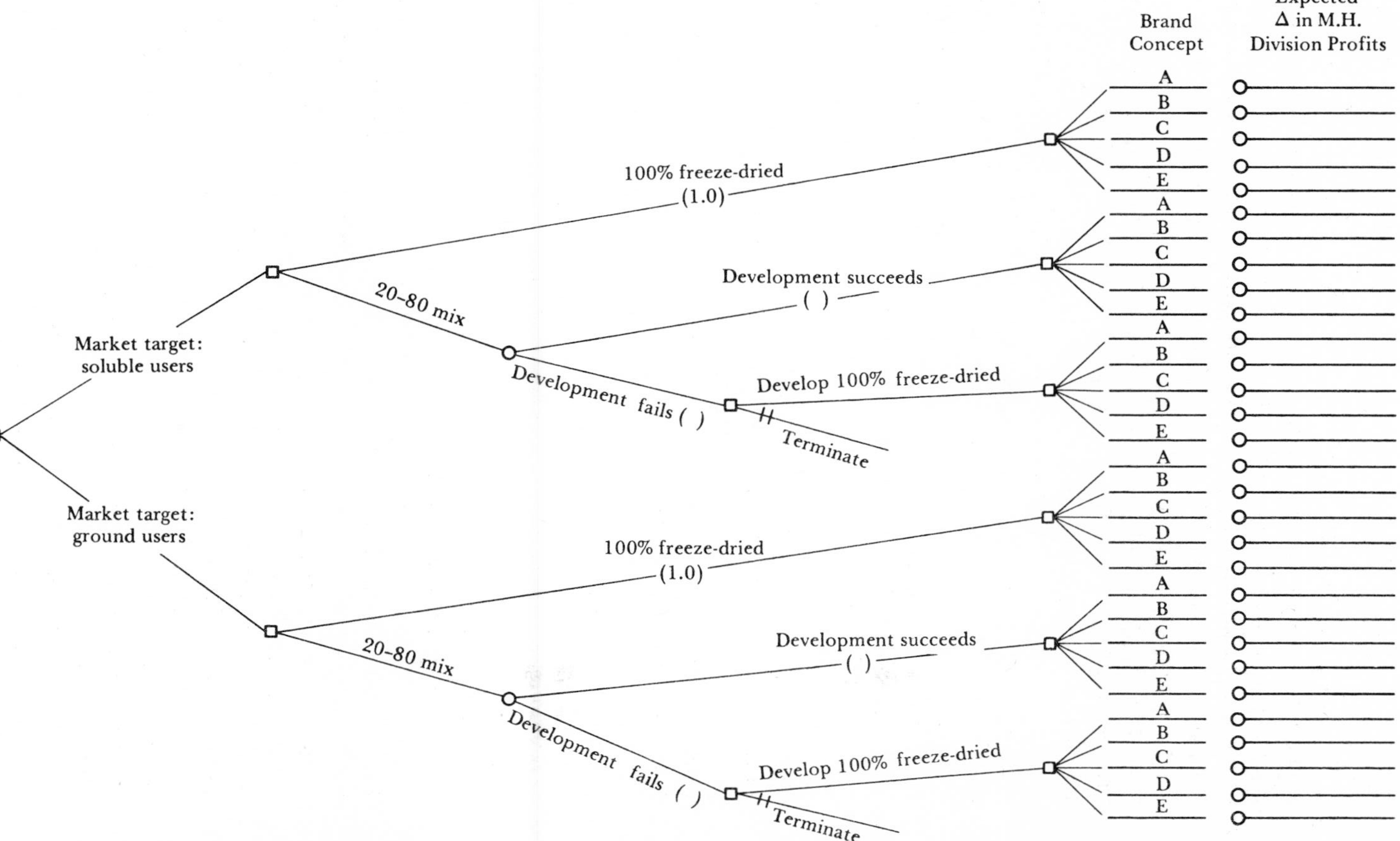

The value of that decision, of course, would be affected by the delay caused by the failure. Termination of the freeze-dried project at that point also would be possible, but this action is ruled out by the same reasoning that led to the initial decision to proceed with freeze-dried-coffee development.

A less-formal way of handling the difference in probability of developmental success would be to carry out all computations as if this probability were the same for all alternative courses of action and then subjectively discount the expected value of the 20–80 mix branches later.

The payoffs of the alternatives, expressed in terms of change in the Maxwell House Division's profit, would appear at the ends of the branches. They would be the net results of a number of judgments and calculations that have not yet been fully specified.

START ON PREPARING FOR FORMAL ANALYSIS

By this time, the reader should be impressed with the complexity of the problem, the need for a systematic procedure for handling it, and the necessity of becoming much more specific. In order to compare the alternatives, it is necessary to specify the measures to be used, the calculations that must be made, and the requisite input data.

Acceptance of the prior decision to proceed with the development of freeze-dried coffee means that attention can be focused on the relative merits of alternative actions rather than their absolute values. The calculations to be made, then, are those necessary to arrive at relative measures of consequences. For the moment, let us confine our attention to relative measures of change in profits of the Maxwell House Division.

In order for the analysis to proceed, a time period must be defined. In this case, it will be governed to a large extent by the lead time the Maxwell House Division is expected to have over competitors with respect to freeze-dried coffee. In order to identify other inputs needed and to think further about a suitable structure for the formal analysis, it is useful at this stage to specify the main steps required for the calculation of the change in profits that would result under a given alternative course of action. In the interest of simplicity, we shall exclude from immediate consideration taxes and present values, for which provisions ultimately must be made.

The following computations would be needed to produce the change in Division profits:

$\triangle PBT$ = gross margin in dollars of freeze-dried brand − (marketing costs + expected increase in plant costs + loss of profits due to cannibalism of sales of other Division brands + cost of any phasing out of other Division brands)

Given this general statement, we can proceed to outline how its elements can be obtained. This is done in terms of the formulas that appear below:

Gross margin = sales in units × gross margin/unit
(Sales in units = sales during market introduction + sum of sales in each going (postintroduction) year)
(Gross margin/unit = selling price − cost of goods sold)
Marketing costs = introductory marketing costs + sum of marketing costs for each going year

(Costs of test marketing need not be computed if they are assumed to be about the same for all alternatives.)

Increase in plant costs = Cost of new plant × annual rate of depreciation × number of years

(If declining-balance rather than straight-line method of depreciation were used, the formula for each year's depreciation would be cost of new plant − depreciation previously taken × depreciation rate.)

(Number of going years = total time period of analysis − (development time + market-test time + market-introduction time))
Loss of profits = cost of phasing out other brand(s) + cost of cannibalism of sales of remaining Division brands
(Cost of phasing out an existing brand = brand's lost sales volume in units × (gross margin/unit − marketing cost/unit))
Cost of cannibalism = (annual cannibalism rate applicable to Division ground-coffee brands × annual ground-coffee volume in units × gross margin/unit × number of years) + (annual cannibalism rate applicable to Division soluble-coffee brands × annual soluble-coffee volume in units × gross margin/unit × number of years

The model just outlined is geared to the decision criterion of change in profits before taxes. Other criteria will be discussed in Chapter 9. The statement of the model does not detail all the calculations needed, but it goes far enough to introduce the reader to the planning of the prior analysis and to permit him to identify required inputs so that work can proceed on arriving at their values. Some of the values can be obtained fairly easily. Others require careful exercise of managerial judgment under uncertainty, the latter to be reflected in the form of probability distributions. There should be several estimates representing the range of possible values for both sales volume and gross margin per unit, for example. The estimates in a given set will be multiplied by their probabilities to get an expected value. Marketing costs might be handled in the same way, although they also could be made conditional on sales volume. Working with the information presented in Chapter 6 and with his own judgment, the reader is encouraged to go as far as he can in determining the basic inputs he would use without actually making the calculations indicated by the above formulas.

The discussion thus far has been carried on under the simplifying assumption that only one brand concept would be used for freeze-dried coffee. The assumption was introduced to make it easier for the reader to begin work on the problem. With the abandonment of this assumption, the problem becomes more complicated but the nature of the approach does not change. Before going further, the reader should begin to consider the problem under an assumption that two brand concepts would be developed. In addition to the task of identifying the two brand concepts to be used, it now becomes necessary to determine which one should be offered first and when the second one should be introduced.

A complex decision problem of major importance should usually be programmed for analysis by computer because of the computational burden involved and the desirability of being able to quickly test the sensitivity of the output to changes in values of the different inputs. A more elaborate model for analysis by computer will be presented in Chapter 9; additional pertinent information will be given in Chapter 8.

STUDY QUESTIONS

1. Without conducting a formal analysis, can you reach a decision you would be willing to act on concerning whether the first freeze-dried coffee should be developed to appeal to users of ground coffee, users of instant coffee, or users of both? Explain your answer.

2. Without conducting a formal analysis, can you reach a decision you would be willing to act on concerning whether the first developmental effort should be directed to 100-percent freeze-dried coffee or to the 20–80 mix? Explain your answer.

3. Assume for the moment that only one freeze-dried-coffee brand is to be marketed by the Maxwell House Division. Which three brand concepts do you regard as most promising and, therefore, perferred subjects for formal analysis? Why?

4. Assume that two brands of freeze-dried coffee are to be marketed by the Maxwell House Division. Identify three two-brand combinations that you regard as the most promising, and explain your reasons for their selection. If only one brand can be introduced at a time, which brand in each of your three two-brand combinations should be marketed first?

CHAPTER 8

DEVELOPING A MODEL FOR THE ANALYSIS OF ONE- AND TWO-BRAND ALTERNATIVES

This chapter supplements Chapter 6, giving additional information that is relevant to the design of a model for computer evaluation of the alternative courses of action on freeze-dried coffee and the requisite input data. The reader should use this new information to continue his own work on model development. Those who plan no such further work should read this chapter as background for understanding the model that will be described in Chapter 9.

The simplifying assumptions temporarily imposed in Chapter 7 to facilitate early thinking are now abandoned. The model to be developed is to be suitable for analyzing two-brand sequences. Its output may consist of more than one kind of measurement for use in comparing the alternatives. Taxes and present values of future flows of funds should be taken into account.

Management, of course, already has decided to go ahead with the freeze-dried-coffee project. We shall assume that the decision is to proceed as rapidly as possible with the development and marketing of two brand concepts, subject to laboratory and production limitations. The new model must recognize several other practical constraints and managerial judgments that will be described shortly. For purposes of illustration, we shall limit our focus in certain ways to simplify handling of the problem without neglecting its important elements.

Management also has decided that freeze-dried-coffee development and marketing should be aimed at the users of ground coffee. It was believed that ground-coffee drinkers, particularly in the West, offered the largest potential for incremental Division sales. The "soluble-coffee stigma" was recognized as a major obstacle to reaching the ground-coffee user. It was decided, therefore, not to become so narrowly oriented to the ground-coffee market that potential customers currently using soluble coffee would be alienated. One Maxwell House Division executive thought that in spite of the planned ground-user orientation, as much as 75 percent of the sales of a freeze-dried product would come from competitive soluble-coffee volume.

A decision had to be made on the mix to be used. Developmental efforts could be concentrated on the preparation of a 100-percent freeze-dried product or on any of a number of possible mixes of freeze-dried and spray-dried coffees. In the interest of simplicity, we shall assume that the mix decision consists only of choosing between these two alternatives: (1) a 100-percent freeze-dried coffee and (2) a mix consisting of 20-percent freeze-dried and 80-percent spray-dried coffees. The same mix would be used in the first two brand concepts to be developed.

THE BRAND ALTERNATIVES

Preliminary screening of the many possible brand-concept alternatives has taken place. To simplify our example, we shall limit our attention to five alternatives to which management wanted to give serious consideration. They are as follows: (1) a Maxwell House freeze-dried coffee replacing the present Instant Maxwell House formulation, (2) a Yuban freeze-dried coffee replacing the present Instant Yuban formulation, (3) a Super Yuban freeze-dried coffee that would be offered along with the present Instant Yuban product, (4) a new brand (with caffein), and (5) a new decaffeinated brand to be marketed in addition to the present Instant Sanka product.

Tentative plans were that the marketing program for either the Maxwell House or Yuban freeze-dried coffees would be directed toward approximately the same market segments presently served by those brands, with a heavy emphasis on attracting new users from the ground-coffee ranks. A freeze-dried product under a new brand would be positioned as a premium coffee offering the taste characteristics of ground coffee and the convenience of soluble coffee. A decaffeinated freeze-dried product would be sold under a new brand without replacing Sanka. A long-run goal, should the development program prove successful, might be to market a freeze-dried product under several brands.

The brand concepts implied somewhat different sets of objectives in terms of desired taste-test results, product cost, product color, and particle size; therefore, they called for different development programs. A decision on priority was necessary.

DEVELOPMENT AND PRODUCTION LIMITATIONS

Limitations of technical manpower and laboratory facilities precluded full simultaneous development of more than one brand concept and mix. Experience gained on the first developmental project would benefit the next one. Any attempt to split resources to meet more than one product-development goal at one time, however, would unduly extend the time period required for obtaining reasonably firm evaluations.

The 100-unit hourly output capacity of the freeze-dried-coffee plant under construction would sustain a test-marketing program for only one brand at a time. Expansion of plant capacity seemed out of the question until the freeze-dry process had given some concrete evidence of success.

The above limitations applied primarily to the first two brand concepts to be chosen. Once they had been developed, accumulated experience and expansion of facilities probably would permit simultaneous development of additional concepts. A decision now on priorities would not constitute an irrevocable commitment to one course of action. The direction could be changed as new findings were obtained. The initial decision, however, was far from trivial because misdirection could waste valuable lead time and resources.

The following sections give estimates of the time requirements and economic factors associated with alternative brand concepts and mixes.[1]

TIME REQUIREMENTS

The time estimates for laboratory and process development, market testing, and introduction to the national market for the five brand concepts are as follows:

	Time Requirements (Years)			
			Market Introduction	
Brand Concept	Laboratory and Process Development	Market Testing	100-percent Freeze-Dried	20–80 Mix
Maxwell House	2½	1	4	3
Yuban	2	1	2½	2
Super Yuban	2	1	2½	2
New brand	1½	1	2½	2
New brand (decaffeinated)	2	1	2½	2

The laboratory and process development time estimates would apply if the brand were given first priority. A brand given second or later priority would require less time because of the experience gained in

[1] Figures are disguised.

the first effort. If the Yuban concept were developed first, for example, about two years would be required to prepare the product for market testing. If Yuban were followed by Maxwell House, the time required for preparing the Maxwell House product for market testing would be considerably shorter, perhaps only one year. The time estimates reflect anticipated differences in developmental difficulty. Given the time estimates, the probability of developmental success was regarded as being approximately the same for the various concepts for any given mix. Differences in development costs were not thought to be large enough for inclusion in the consideration.

Regardless of the order of development, about one year would be required for test marketing each brand after product development had been completed. Test-marketing costs would be about the same for the different brands.

If positive results were obtained in the market tests, national market introduction would commence. The listed market-introduction times are estimates of how long it would take to achieve national distribution after test marketing had been completed. They apply regardless of brand order, and they assume favorable results all along the way. They take into account the speed with which it would be desirable and possible to build new freeze-dry plants. The shorter time estimate for the 20–80 mix reflects its lower requirement of freeze-dried-coffee output.

SALES VOLUMES, GROSS MARGINS, AND MARKETING COSTS

Sales volumes and gross margins for freeze-dried-coffee brands would depend, of course, on the price charged. Average selling prices, gross margins, and marketing costs experienced by the Maxwell House Division's spray-dried-coffee brands are given in Exhibit 8–1. Estimates of the gross margins for freeze-dried coffee priced at different premiums over the price of the Division's spray-dried brands are given in Exhibit 8–2. They assume the use of 100-percent freeze-dried coffee. If the 20–80 mix were used, the gross margins would be about 25 percent higher for all brands.

Annual sales volumes would have to be estimated for each freeze-dried brand under consideration. The figures would vary depending on whether the brand would be the first or the second to be marketed by the Maxwell House Division. They would also vary depending on the extent and timing of freeze-dried-coffee competition and the nature of the competitive brand concepts. Buildup of sales to *going-year* volume (that is, after market introduction had been completed) would be approximately linear with time.

Should development and marketing of freeze-dried coffee prove feasible, it was expected that Maxwell House Division coffee sales would be first affected by freeze-dried-coffee competition from at least one major company by January 1, 1968, with a probability of .25; by January 1, 1969, with a probability of .45; and by January 1, 1970, with a probability of .30.

Marketing costs would vary by brand and sales volume. (See Exhibit 8–3 for preliminary estimates made by the product manager.) Introductory marketing costs, incremental to going-year marketing costs, would be spread about evenly across the introductory years for any one brand. They were estimated as follows: (1) Maxwell House, about $21 million; (2) Yuban, about $9 million; (3) Super Yuban, about $6 million; (4) a new brand with caffein, about $13 million; and (5) a new decaffeinated brand, about $17 million.

General and administrative expenses other than marketing costs and depreciation charges were not expected to differ significantly on the basis of priorities given to the brand concepts and mixes.

CANNIBALISM

Cannibalism—the switching of sales from existing Maxwell House Division brands to a freeze-dried brand—was expected to be a significant factor. For Maxwell House and Yuban freeze-dried coffees, it was assumed that all sales up to a base volume (20 million units for Maxwell House and 3 million units for Yuban) would represent no net increase in Division sales. Instead, they would represent sales volume lost by the phasing out of the Maxwell House and Yuban spray-dried coffees. The gross margins and marketing costs for the two spray-dried coffees were $3.00 and $1.50 per unit for Maxwell House and $3.25 and $1.75 per unit for Yuban. Sales in excess of the base sales volumes were expected to involve cannibalism.

Preliminary estimates of cannibalism rates made by the product manager follow: (1) about 20 percent of the Maxwell House freeze-dried-coffee sales over base volume would come from Division ground-coffee sales and 10 percent from other Division soluble-coffee sales; (2) about 10 percent of the Yuban freeze-dried-coffee sales over base volume would come from Division ground-coffee sales and 15 percent from Division soluble-coffee sales; (3) about 15 percent of Super Yuban freeze-dried-coffee sales would come from Division ground-coffee sales and 30 percent from Division soluble sales; (4) about 10 percent of the sales of a new freeze-dried brand (with caffein) would come from Division ground-coffee sales and 25 percent from Division soluble-coffee sales; (5) about 15 percent of a new decaffeinated freeze-dried brand's sales would come from Division ground-coffee sales and 35 percent from Division soluble-coffee sales.

The estimates just given assumed that the brand would be the first freeze-dried coffee marketed. Cannibalism rates for second brands remained to be estimated. The rates would vary depending on the identity of the first freeze-dried brand.

ESTIMATES RELEVANT TO THE CHOICE OF MIX

On the basis of information collected to date, the probability that the 20–80 mix could be developed to meet product-formulation objectives was judged to be about one-fourth of that for the 100-percent freeze-dried product. There were significant economic advantages to the 20–80 mix, however, that made it worthy of consideration.

New plant and equipment for the 100-percent freeze-dried product were expected to cost about $5 million per million units of annual volume for the Maxwell House, Yuban, Super Yuban, and new brand (with caffein) alternatives and about $5.5 million per million units for a new decaffeinated brand. If the 20–80 mix were used, the plant costs would be about one-third of the figures just given. Depreciation charges for the purpose of evaluation of the alternatives could be based on a useful life of ten years and computed by the declining-balance method. (The gross-margin figures given in Exhibit 8–2 do not take into account the costs of new plant and equipment.)

In addition to its lower investment requirement, the 20–80 product would be somewhat cheaper to make. The gross-margin estimates given in Exhibit 8–2 should be increased 25 percent if the brands were made with the 20–80 mix.

NEEDED: A DECISION ON PRIORITIES

The information given in this chapter and in Chapter 6 provides a basis for planning in detail a formal prior analysis of the alternatives preliminary to deciding on product development and marketing priorities. A suitable recommendation for management would consist of specifying the first two brand concepts for development, the order in which they should be developed, and the mix alternative to be used. The model for making these decisions should be outlined, and assumptions for use in the analysis should be specified.

STUDY QUESTIONS

A general approach to the prior analysis of the freeze-dried-coffee problem was indicated in Chapter 7. The questions that follow are given for the benefit of those who would like to go more deeply into model construction.

1. What model should be used to determine which two of the alternative brand concepts should be developed first, the order of

their development, and the mix alternative that should be adopted?

a. How complex should the model be?

b. What simplifying assumptions should be made?

c. What would be the impact of the simplifying assumptions on the analysis?

2. Prepare a decision tree diagram of the basic structure of your model for the prior analysis.

EXHIBIT 8–1

AVERAGE SELLING PRICES, GROSS MARGINS, AND MARKETING EXPENSES FOR INSTANT MAXWELL HOUSE, INSTANT YUBAN, AND INSTANT SANKA DURING PAST YEAR [a]

(Per Unit Except Where Otherwise Specified)

	Instant Maxwell House	Instant Yuban	Instant Sanka
Retail price	$8.00	$8.97	$9.50
Trade margin (10%)	.80	.82	.95
Manufacturer's selling price	$7.20	$8.15	$8.65
Cost of goods sold	4.20	4.90	5.40
Gross margin	3.00	3.25	3.25
Marketing expenses	1.50	1.75 [b]	1.75
Contribution to other expenses and profit	$1.50	$1.50	$1.50
Cost of Goods Sold			
Green coffee	$3.20	$3.70	$3.25
Packaging materials	.50	.60	.55
Processing	.50	.60	1.60
	$4.20	$4.90	$5.40
Marketing Expenses			
Advertising media	$ .35	$ .75	$ .90
Promotional deals	.73	.65	.50
Field selling and other	.42	.35	.40
	$1.50	$1.75	$1.80
Average Price Per Ounce			
Retail price	$ .165	$ .187	$ .198
Manufacturer's selling price	.150	.170	.180
Estimated retail cost per cup of coffee, at 13 cups per ounce	$ .0127	$ .0144	$ .0152
Total units sold (millions)	21.0	2.2	4.5

[a] Disguised estimates. Figures are weighted averages which recognize that each brand was sold in three different package sizes: Instant Maxwell House, in two-, six-, and ten-ounce jars; Instant Yuban, in two-, five-, and nine-ounce jars; and Instant Sanka, in two-, five-, and eight-ounce jars.

[b] Expected "normal" marketing expenses. Actual expenses for Instant Yuban in past year were abnormally high because of market expansion efforts. Instant Yuban's sales in the next year were expected to be about three million units.

EXHIBIT 8–2

ESTIMATED SELLING PRICES AND GROSS MARGINS AT DIFFERENT PRICE PREMIUMS OVER PRICES OF CURRENT BRANDS

	Per Ounce		Per Unit [a]		
Percent Premium	Average Retail Price	Average Manu-facturer's Price	Average Manu-facturer's Price	Cost of Goods Sold [b]	Gross Margin [c]
At Premiums Over Price of Instant Maxwell House					
0	$0.165	$0.150	$ 7.20	$5.20	$2.00
10	.182	.165	7.92	5.20	3.08
20	.198	.180	8.64	5.20	3.80
30	.215	.195	9.36	5.20	4.52
40	.231	.210	10.08	5.20	5.24
At Premiums Over Price of Instant Yuban					
0	.187	.170	8.15	5.90	2.25
10	.206	.187	8.97	5.90	3.07
20	.224	.204	9.78	5.90	3.88
30	.243	.221	10.60	5.90	4.70
40	.262	.238	11.41	5.90	5.51
At Premiums Over Price of Instant Sanka					
0	.198	.180	8.65	6.40	2.25
10	.218	.198	9.52	6.40	3.12
20	.238	.216	10.38	6.40	3.98
30	.257	.234	11.25	6.40	4.85
40	.277	.252	12.11	6.40	5.71

[a] One unit equals forty-eight ounces of soluble coffee.

[b] Cost of goods sold is based on the assumption that freeze-dried coffee ultimately could be produced at a cost of $1.00 more than that incurred for the current spray-dried-coffee brands. Early production costs were expected to exceed the costs for the spray-dried brands by $1.50 per unit.

[c] The gross-margin estimates are for a 100-percent freeze dried coffee. They should be increased by about 25 percent for all brands when considering the 20–80 mix.

EXHIBIT 8–3

ESTIMATED MARKETING COSTS, BY FREEZE-DRIED-COFFEE BRANDS, FOR DIFFERENT LEVELS OF SALES VOLUME
(In Dollars Per Unit)

Brand	Sales-Volume Estimate [a] Highest	2d Highest	3d Highest	Lowest
Maxwell House	1.40	1.55	1.70	1.80
Yuban	1.65	1.70	1.80	2.00
Super Yuban	1.50	1.55	1.70	1.80
New brand (caffein)	1.55	1.60	1.70	1.90
New brand (decaffeinated)	1.55	1.60	1.75	1.95

[a] The four levels for each brand represent the range of its possible sales volume as seen by management. The sales figures are not given so that the reader can develop his own estimates without being biased by the judgments of someone else.

CHAPTER 9

EVALUATING ONE- AND TWO-BRAND ALTERNATIVES USING THE COMPUTER

In order to illustrate the procedure for undertaking a prior analysis of a complex choice problem, this chapter presents a model for evaluating the one- and two-brand alternatives represented by the five freeze-dried-coffee brand concepts described in Chapter 8.

The material is organized to make it easy for the reader to examine the model briefly so that he may become generally familiar with its principal features. A decision tree showing the brand-mix alternatives is presented in Figure 9–1 in order to provide an overview. This is supplemented in the text by an outline of the main steps for calculating the end values for the tree and the expected consequences for each course of action.

For those who wish to go into the assumptions and procedures of the model in depth, complete flow diagrams, accompanied by written descriptions, are available from Harper & Row, Publishers, in a companion publication, *Computer Programs and Flow Diagrams for Management Applications of Decision Theory*. The same volume contains the computer program itself and definitions of the variables used. For the convenience of students, instructions for punching input data cards for making use of the program are included in Exhibit 9–2. The materials are valuable for persons who would like to experiment with using the computer in the classroom handling of management problems. Subject to the assumptions built into the model, the program can be used to analyze different sets of inputs. Because they are relieved of the computational burden, students can focus on the more rewarding activity of identifying and explaining the main differences in outcomes and examining the underlying assumptions that account for them.

THE CONTEXT

The freeze-dried-coffee decision problem was described at length in order to allow the reader to assume vicariously the role of a manager who must evaluate a situation and come up with specific supportable recommendations on the basis of available information. The reader was encouraged to go as far as he could in structuring the problem for analysis and identifying needed input data. This approach was taken as a means of cultivating understanding of the kinds of judgments required, the difficulty of making them under uncertainty, and the procedures that can be helpful.

In view of the complexity of the problem, an explicit quantitative model appears necessary for the careful consideration of the various factors on which the outcome of a course of action depends and the interactions of these factors. The problem can be modeled at different degrees of complexity, depending on the number and magnitude of the simplifying assumptions one wishes to make. The larger the number and the magnitude of the assumptions, the simpler the model and, frequently, the lower the probability that the model is adequately representing the problem. If one were doing the analysis with pencil and slide rule, a simple model probably would be chosen. The computer makes feasible the use of more complete models, but, of course, it does not preclude the use of important assumptions. The model described here is a relatively fully developed one although, as will be discussed later, it could be elaborated in a number of different ways.

THE MODEL

Inasmuch as management already had decided to attempt the development of freeze-dried coffee, the model takes this decision as given and focuses on the choice of a course of action for implementing it. It does so by providing for the calculation of relative measures of thirty-six of the forty two-brand alternatives represented by the five brands and two mixes under consideration (eighteen two-brand sequences for each mix). The Yuban–Super Yuban and Super Yuban–Yuban sequences were eliminated because the brands were seen as being too directly competitive for the second brand to produce satisfactory incremental sales and profits.

The flow of the analysis is indicated by the decision tree in Figure 9–1. The first choice represented is for the mix for the initial development attempt. If developmental efforts for the 20–80 mix failed, they would

FIGURE 9-1

DECISION TREE FOR ANALYSIS OF TWO-BRAND SEQUENCES FOR FREEZE-DRIED COFFEE[1]

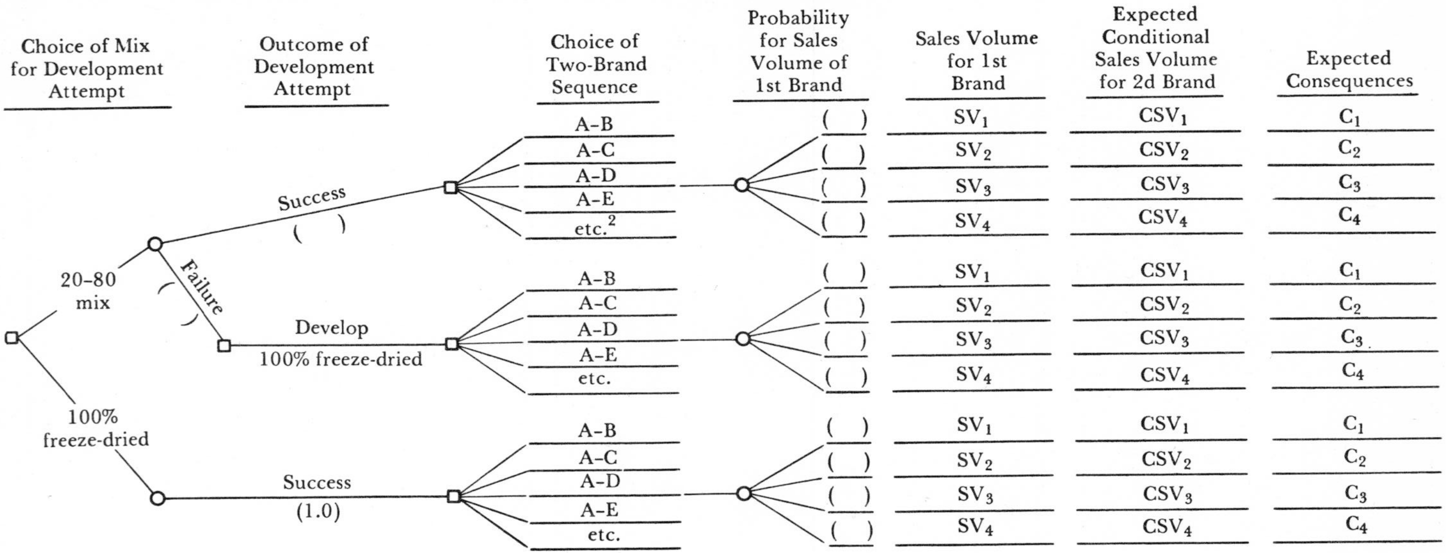

Choice of Mix for Development Attempt
Outcome of Development Attempt
Choice of Two-Brand Sequence
Probability for Sales Volume of 1st Brand
Sales Volume for 1st Brand
Expected Conditional Sales Volume for 2d Brand
Expected Consequences

20–80 mix
Success ( )
Failure ( )
Develop 100% freeze-dried
100% freeze-dried
Success (1.0)

Choice of Two-Brand Sequence	Probability for Sales Volume of 1st Brand	Sales Volume for 1st Brand	Expected Conditional Sales Volume for 2d Brand	Expected Consequences
A–B	( )	SV_1	CSV_1	C_1
A–C	( )	SV_2	CSV_2	C_2
A–D	( )	SV_3	CSV_3	C_3
A–E	( )	SV_4	CSV_4	C_4
etc.[2]				
A–B	( )	SV_1	CSV_1	C_1
A–C	( )	SV_2	CSV_2	C_2
A–D	( )	SV_3	CSV_3	C_3
A–E	( )	SV_4	CSV_4	C_4
etc.				
A–B	( )	SV_1	CSV_1	C_1
A–C	( )	SV_2	CSV_2	C_2
A–D	( )	SV_3	CSV_3	C_3
A–E	( )	SV_4	CSV_4	C_4
etc.				

[1] The diagram is completed for the A–D brand sequence only because of space limitations.
[2] The remainder of the possible two-brand sequences are omitted from the diagram because of space limitations.

be switched immediately to the 100-percent freeze-dried product, for which success is assumed. The developmental paths specifying the mix as well as the brand concepts have been set up as being mutually exclusive. The assumption was made that both brands in a given two-brand sequence would use the same mix because of the constraints on the development program.

Calculations of the measures that are to be the output of the analysis are based on sales-volume estimates for the second brand that are conditional on the identity and the sales volume of the first brand. (The decision tree in Figure 9–1 is completed only for the A–D brand sequence because of space limitations).

The model provides for the output of several different measures of the consequences of a course of action.[1] This is done to illustrate the measures and to facilitate later discussion of the choice and handling of multiple decision criteria. The measures are as follows:

1. *Accumulated Net Present Value of Expected Cash Flow.* This measure reflects both the size and the timing of expected cash receipts and expenditures. In its computation, the expected cash flow (cash inflow minus cash outflow) for each year of the analysis is discounted at a rate chosen to represent what would be lost by receiving the money later rather than now. The algebraic sum of the resulting present values of the yearly cash flows is then computed.

2. *Expected Internal Rate of Return on Investment.* This is the rate at which an investment is repaid by the proceeds. More specifically, it is the rate that equates the present value of the expected cash inflows with the expected cash outflows. An equivalent statement is that it is the rate of interest that makes the net present value of cash flow zero. The measure sometimes is referred to as the *marginal efficiency of investment.*

3. *Profitability Index.* This measure is simply the ratio of the accumulated present value of the expected net cash inflows from operations to the present value of the expected investment outlays.

4. *Exposure to Gain and Loss.* For each course of action, several values that are representative of the range of the possible consequences are given, along with their probabilities of occurrence.

[1] In considering appropriate investment criteria, the reader may wish to consult references on capital budgeting such as the following: Harold Bierman, Jr., and Seymour Smidt, *The Capital Budgeting Decision* (New York: The Macmillan Company, 1966); James C. Van Horne, *Financial Management and Policy* (Englewood Cliffs, N.J.: Prentice-Hall, Inc., 1968), pp. 15–59; William J. Baumol, *Economic Theory and Operations Analysis* (Englewood Cliffs, N.J.: Prentice-Hall, Inc., 1965), pp. 434–475.

Consequences are expressed as accumulated net present values of expected cash flow.

All four measures are net of federal income taxes. The same measures can be produced for ten alternative courses of action calling for the development of a single-brand concept using either 100-percent freeze-dried coffee or the 20–80 mix.

The measures will be relative rather than absolute because they will not reflect costs for which variations by alternative were judged to be small enough to be disregarded for the decision purpose at hand. Items to be omitted from consideration on these grounds include product-development cost, the cost of the pilot plant, test-marketing costs, and general and administrative expenses.

Before going into greater detail, the procedure for calculating an expected value for a course of action, given the information identified in the decision tree, will be illustrated. (The flow of calculations for the tree is from right to left.) For example, assume that the appropriate figures for accumulated present value of expected cash flow for the A–D brand sequence were in the Expected Consequences column of the tree in place of their symbols (C_1, C_2, C_3, and so forth). (We will show where the values come from shortly.)

The overall accumulated present value of expected cash flow for a decision to direct first developmental efforts to 100-percent freeze-dried coffee and adopt the A–D brand sequence would be computed by multiplying the accumulated present value of expected cash flow associated with each of the four levels of sales volume by their respective probabilities and summing the products. The result would be the expected value for the course of action just identified because of the assumption of certainty for developmental success.

The overall accumulated present value of expected cash flow for a decision to direct first developmental efforts to the 20–80 mix and adopt the A–D brand sequence would involve the following steps: (1) multiply the accumulated present value of expected cash flow associated with each of the four levels of sales volume under the assumption of successful development of the 20–80 mix by its respective probability and sum the products; (2) multiply the result of step 1 by the probability of successful development of the 20–80 mix; (3) multiply the accumulated present value of expected cash flow associated with each of the four levels of sales volume under the assumption of failure to develop the 20–80 mix followed by successful development of the 100-percent freeze-dried product by its respective probability and sum the products; (4) multiply the result of step 3 by the probability of developmental failure for the 20–80 mix; and (5) sum the products obtained in steps 2 and 4. The expected values for the two courses of action can be considered, along with other relevant factors, in the process of deciding which alternative should be adopted.

An overview of the computer program will be helpful at this point in the discussion. It identifies the main calculations that produce the end values of the decision tree as well as those calculations already mentioned for using the end values to compute overall expected values for a course of action. Steps involved in evaluating a decision to develop the 100-percent freeze-dried product and adopt a given two-brand sequence are outlined. A discussion of points of difference represented by the procedures for evaluating a decision to direct initial developmental efforts to the 20–80 mix and adopt a given two-brand sequence and for evaluating one-brand alternatives follows.

Before examining the outline of the program, the reader will find it helpful to acquaint himself with the inputs required by the model. They are specified in an input data sheet (see Exhibit 9–1) that will become the object of more detailed discussion in Chapter 10. Assignments at the end of the chapter call for the preparation of a set of input data.

As indicated by the outline that follows, the program first provides for certain steps to be taken after the reading-in of the input data before the various values for each year can be calculated. The yearly values are then produced prior to their use in computations of the values for the entire period of analysis.

MAIN STEPS IN ANALYSIS OF A DECISION TO DEVELOP 100-PERCENT FREEZE-DRIED COFFEE AND ADOPT A GIVEN TWO-BRAND SEQUENCE

The steps for the calculation of the various measures that comprise the output are outlined below approximately in the order of their appearance in the computer program. The program assumes that an attempt to develop a 100-percent freeze-dried coffee would be successful.

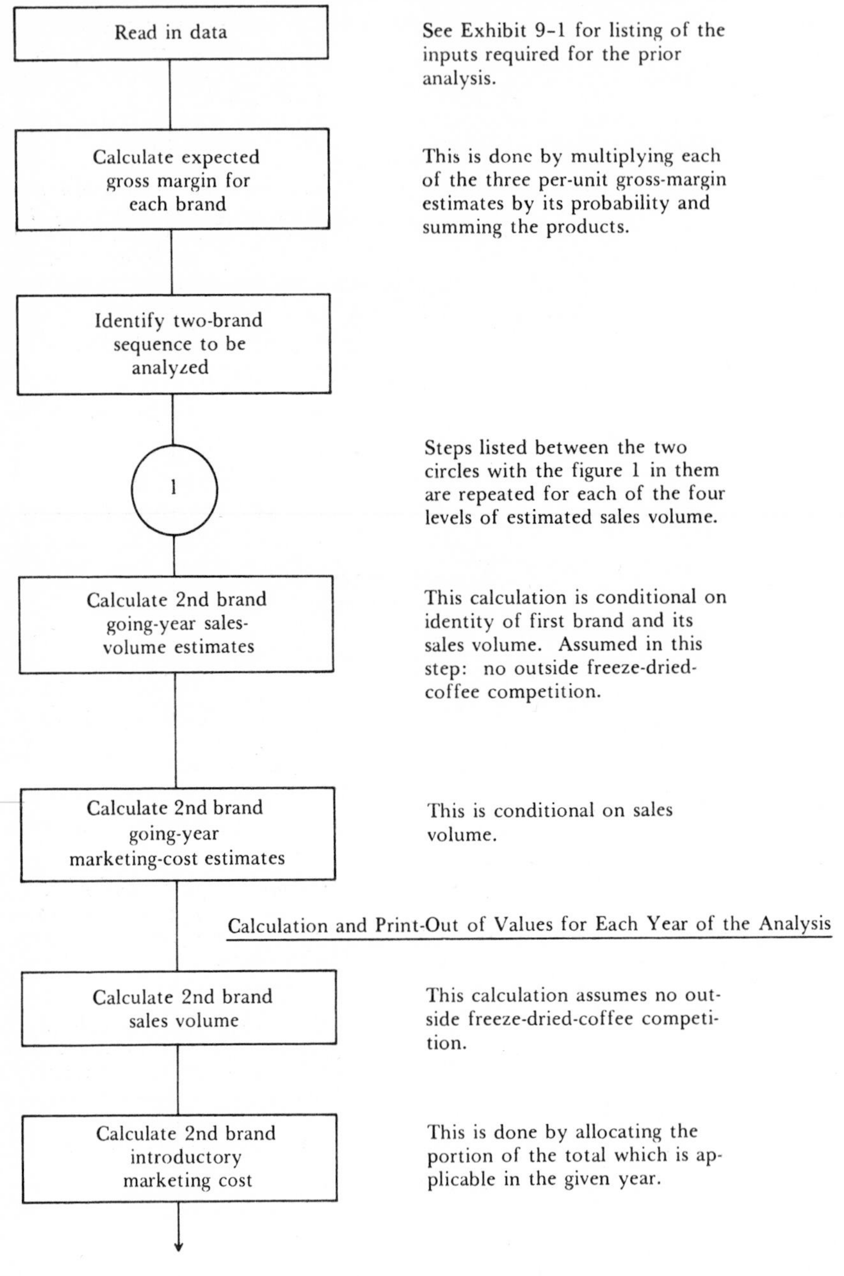

Steps Preliminary to Calculation of Yearly Values
Read in data
See Exhibit 9-1 for listing of the inputs required for the prior analysis.
Calculate expected gross margin for each brand
This is done by multiplying each of the three per-unit gross-margin estimates by its probability and summing the products.
Identify two-brand sequence to be analyzed
1
Steps listed between the two circles with the figure 1 in them are repeated for each of the four levels of estimated sales volume.
Calculate 2nd brand going-year sales-volume estimates
This calculation is conditional on identity of first brand and its sales volume. Assumed in this step: no outside freeze-dried-coffee competition.
Calculate 2nd brand going-year marketing-cost estimates
This is conditional on sales volume.
Calculation and Print-Out of Values for Each Year of the Analysis
Calculate 2nd brand sales volume
This calculation assumes no outside freeze-dried-coffee competition.
Calculate 2nd brand introductory marketing cost
This is done by allocating the portion of the total which is applicable in the given year.

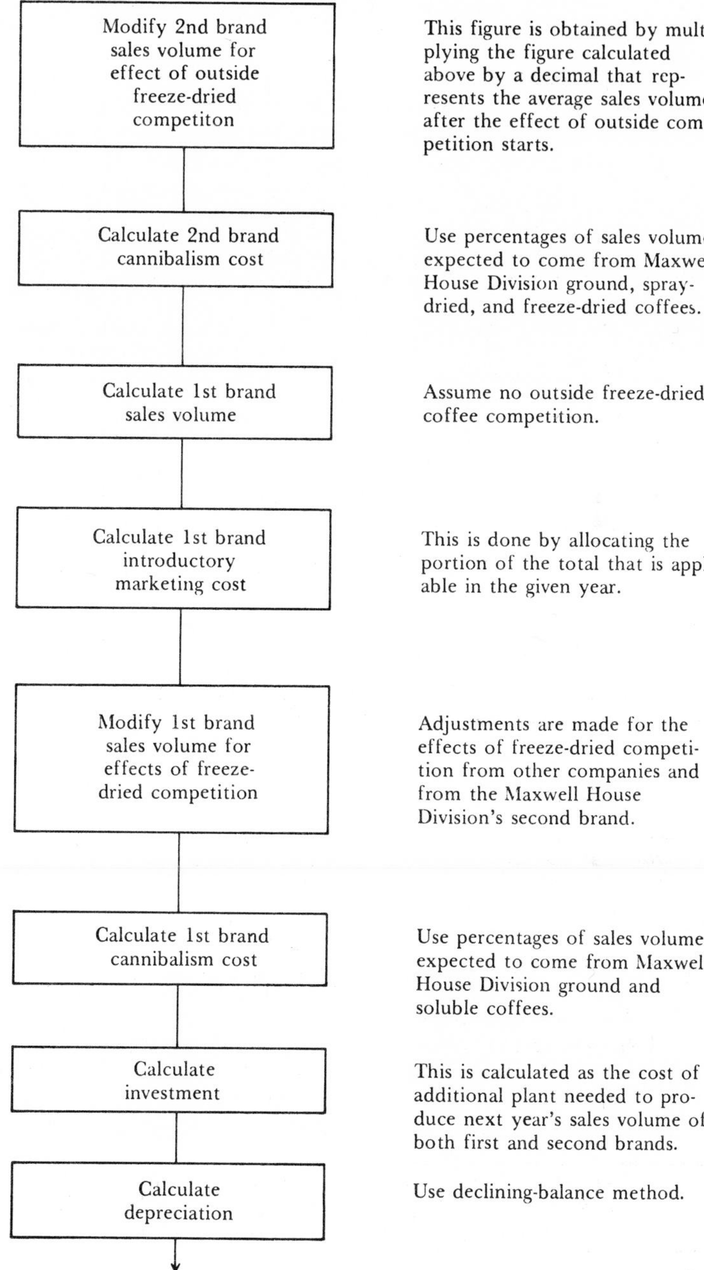
Modify 2nd brand sales volume for effect of outside freeze-dried competiton
This figure is obtained by multiplying the figure calculated above by a decimal that represents the average sales volume after the effect of outside competition starts.
Calculate 2nd brand cannibalism cost
Use percentages of sales volume expected to come from Maxwell House Division ground, spray-dried, and freeze-dried coffees.
Calculate 1st brand sales volume
Assume no outside freeze-dried-coffee competition.
Calculate 1st brand introductory marketing cost
This is done by allocating the portion of the total that is applicable in the given year.
Modify 1st brand sales volume for effects of freeze-dried competition
Adjustments are made for the effects of freeze-dried competition from other companies and from the Maxwell House Division's second brand.
Calculate 1st brand cannibalism cost
Use percentages of sales volume expected to come from Maxwell House Division ground and soluble coffees.
Calculate investment
This is calculated as the cost of additional plant needed to produce next year's sales volume of both first and second brands.
Calculate depreciation
Use declining-balance method.

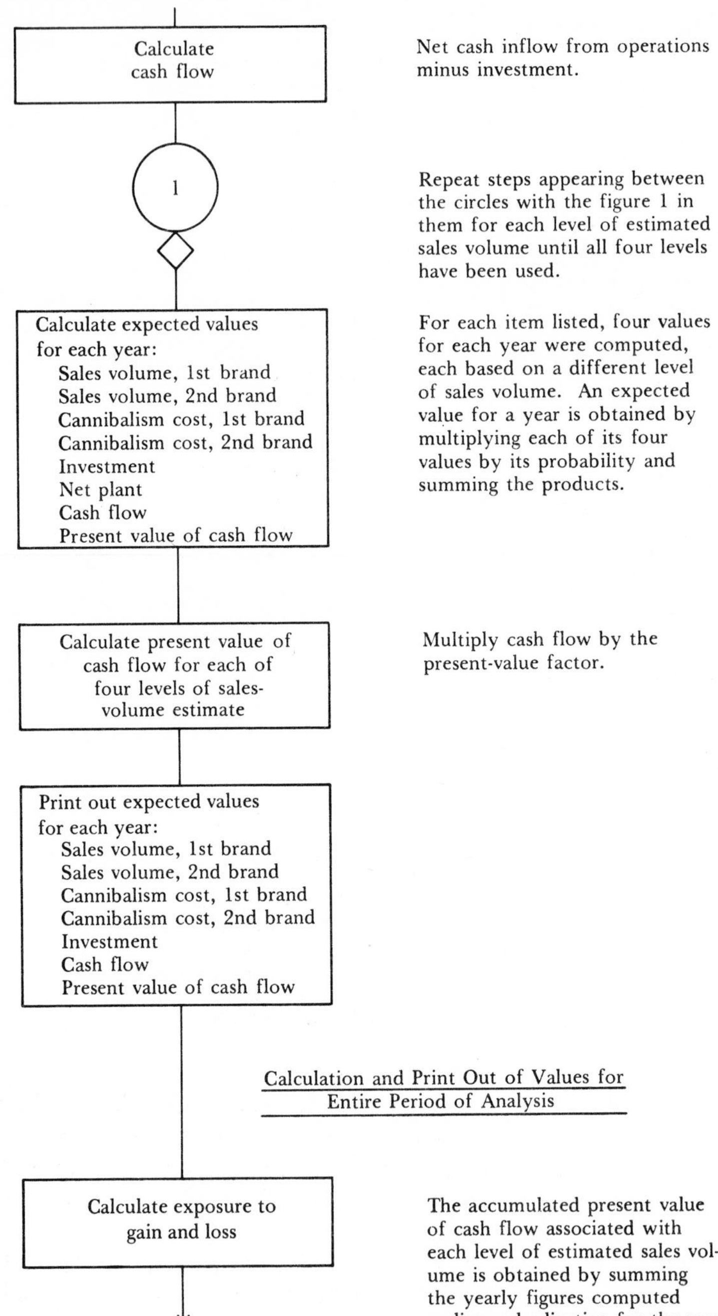

Calculate cash flow
Net cash inflow from operations minus investment.
1
Repeat steps appearing between the circles with the figure 1 in them for each level of estimated sales volume until all four levels have been used.
Calculate expected values for each year:
Sales volume, 1st brand
Sales volume, 2nd brand
Cannibalism cost, 1st brand
Cannibalism cost, 2nd brand
Investment
Net plant
Cash flow
Present value of cash flow
For each item listed, four values for each year were computed, each based on a different level of sales volume. An expected value for a year is obtained by multiplying each of its four values by its probability and summing the products.
Calculate present value of cash flow for each of four levels of sales-volume estimate
Multiply cash flow by the present-value factor.
Print out expected values for each year:
Sales volume, 1st brand
Sales volume, 2nd brand
Cannibalism cost, 1st brand
Cannibalism cost, 2nd brand
Investment
Cash flow
Present value of cash flow
Calculation and Print Out of Values for Entire Period of Analysis
Calculate exposure to gain and loss
The accumulated present value of cash flow associated with each level of estimated sales volume is obtained by summing the yearly figures computed earlier and adjusting for the market value of the plant at the end of the period of analysis.

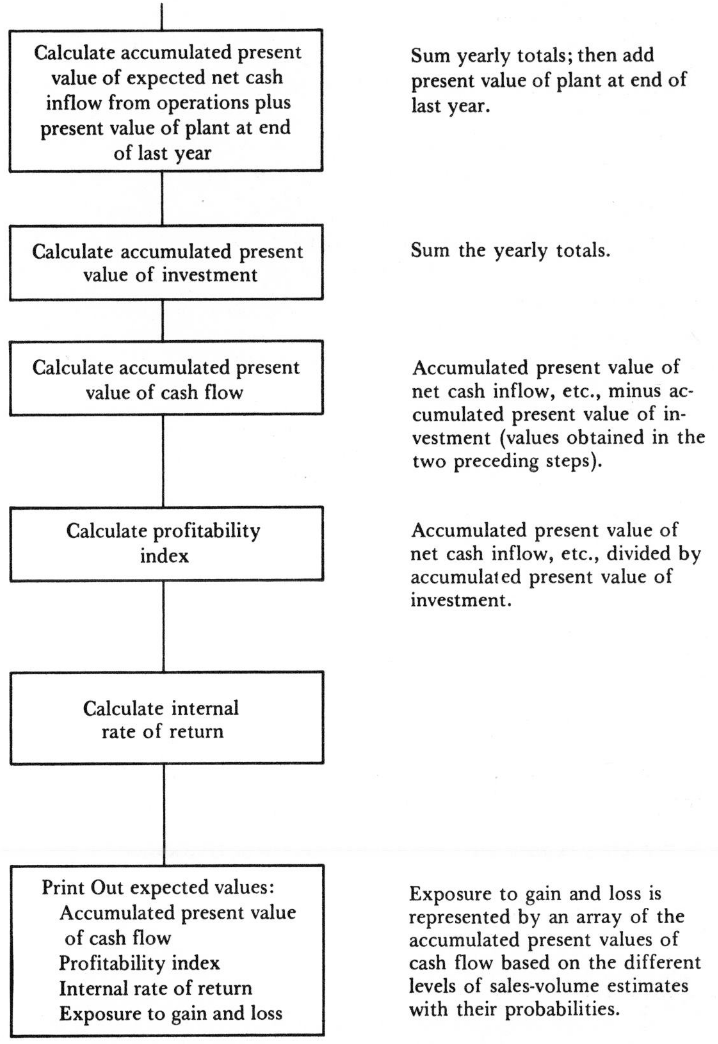
Calculate accumulated present value of expected net cash inflow from operations plus present value of plant at end of last year
Sum yearly totals; then add present value of plant at end of last year.
Calculate accumulated present value of investment
Sum the yearly totals.
Calculate accumulated present value of cash flow
Accumulated present value of net cash inflow, etc., minus accumulated present value of investment (values obtained in the two preceding steps).
Calculate profitability index
Accumulated present value of net cash inflow, etc., divided by accumulated present value of investment.
Calculate internal rate of return
Print Out expected values: Accumulated present value of cash flow Profitability index Internal rate of return Exposure to gain and loss
Exposure to gain and loss is represented by an array of the accumulated present values of cash flow based on the different levels of sales-volume estimates with their probabilities.

Essentially the same steps identified above are followed in the analysis of two-brand alternatives associated with a decision to focus the initial developmental work on the 20–80 mix. More calculations are required, however, because the attempt may either succeed or fail and the output values must reflect the uncertainty.

Two sets of values must be produced, one for success and one for failure. In either case, the yearly values will not be the same as those computed for a decision in favor of the same two-brand sequence and development of 100-percent freeze-dried coffee. Successful development of the 20–80 mix would make possible earlier market introduction and lower production and plant costs. If the first development project were for the 20–80 mix and it failed, it is assumed that efforts would be switched immediately to the 100-percent freeze-dried product and that they would succeed, resulting in a two-year delay in getting a freeze-dried brand to market.

In addition to the computation of the requisite yearly values, the program provides for calculating and printing out the following final values, which allow for the consequences of success and failure on developing the 20–80 mix and their respective probabilities: yearly expected values of cash flow, yearly expected present values of cash flow, accumulated present value of expected cash flow, profitability index, internal rate of return, and exposure to gain and loss.

MAIN STEPS IN ANALYSIS OF ONE-BRAND ALTERNATIVES

Analysis of one-brand alternatives proceeds in the same way as the analysis of two-brand alternatives does, except that the steps representing inputs, calculations, and print-outs for the second brand are omitted.

ASSUMPTIONS OF THE MODEL

The model, of course, employs many assumptions, a number of which have already been mentioned. Inasmuch as they are basic to an understanding of the model, the principal assumptions are identified here and the rationales for their use are explained briefly. The suitability of various assumptions will receive additional attention in Chapter 10.

THE OUTPUT. A basic assumption, of course, is that the four kinds of measures which constitute the output correspond with decision criteria central to management concerns. Another assumption is that an output of relative values is satisfactory for the purpose of choosing among alternatives, given that the decision already had been made to proceed with the freeze-dried-coffee project. The relative values differ from absolute values primarily in that certain costs regarded as being about the same for all alternatives were omitted.

PRESENT VALUES. Also assumed is the appropriateness of the use of present values for comparing the alternatives. The latter would have different financial consequences at different points of time over a number of years. To put them on a comparable basis, the value of the future financial consequences is computed as of the time of the decision by discounting them.

TIME PERIOD. The model provides for time periods of analysis up to fifteen years. It is assumed that some periods of less than fifteen years in length are long enough to permit measurement of all significant change in the relative positions of the alternatives under examination.

TIME SCHEDULE. The model assumes that development and marketing of two freeze-dried-coffee brands should proceed as rapidly as possible, subject to the laboratory and production limitations and the time requirements for development and market testing. Following such a schedule would allow the company to take maximum advantage of its lead time over competition in marketing what management believed to be a significant product innovation. The assumption was applied alike to all alternatives. If the decision were for a slower schedule, the values calculated by the model would be different, but it was assumed that this would not affect the rank order of the alternatives.

DEVELOPMENT TIME. The value of this variable was assumed to be independent of the choice of mix, except that efforts to develop the 20–80 mix would be continued only for a specified length of time before they would be switched to the 100-percent freeze-dried product if they were not successful. The model computes development time for second brands to be half that for first brands, reflecting expected benefits of learning from the first developmental efforts.

PROBABILITY OF DEVELOPMENTAL SUCCESS. The model takes into account the judgment that the probability of developmental success for the 20–80 mix was 25 percent of that for 100-percent freeze-dried coffee and recognizes the opportunity cost of a developmental failure on the 20–80 mix. The model also assumes that it is certain that the 100-percent freeze-dried product can be developed successfully. Both assumptions were consistent with management judgment, and the latter assumption has an advantage of simplifying calculations.

MIX. The assumption is made that the first two brand alternatives would use the same mix, reflecting in part the constraints of the developmental program. If a successful 20–80 mix should result from the first developmental efforts, it would be used for both brands because of its cost advantages. If the 100-percent freeze-dried coffee should be developed first, an attempt might then be made to develop the mix. It was assumed, however, that such an attempt could not succeed before the introduction of the second brand. Once available, the 20–80 mix presumably would be used in all Maxwell House Division freeze-dried-coffee brands.

SALES VOLUMES. The model provides for using the same set of sales-volume estimates for each brand whether the latter were based on the 20–80 mix or on 100-percent freeze-dried coffee. The assumption is that in either case the brand would have to meet the same taste criteria for its development to be judged successful and, therefore, that it would have the same appeal to consumers.

MARKET INTRODUCTION. The model assumes linear growth in sales for each brand-mix alternative until going-year volume is reached. Although the buildup is more likely to occur in a stair-step manner, information on this point was not immediately available. The simplifying linear assumption applied equally to all alternatives seemed reasonable for the objectives of the model.

GROSS MARGIN. Gross margin is assumed to be independent of sales volume within the range visualized for each brand. The assumption reflected management judgment that it was correct enough to warrant its use.

CANNIBALISM COSTS. The model provides for charging cannibalism as a cost before but not after outside freeze-dried-coffee competition is expected to affect Maxwell House Division sales. This was regarded as being consistent with the objective of choosing an alternative that would maximize Division profits. Once outside freeze-dried-coffee competition appeared, the assumption that sales of Division brands would continue at the same levels if the Division itself did not subject them to freeze-dried competition would no longer be valid.

INVESTMENT. The model handles investment as a variable cost incurred one year before the plant capacity is needed. Although the procedure departs from realism to an unknown degree, it was assumed that it would nevertheless produce useful relative values for the alternatives. Ideally, an investment-schedule input for each alternative would be furnished from another analytical model that would consider optimum plant size, the amount of plant capacity that should be built in view of the annual sales forecasts, construction time, and other factors. In the absence of such input data, simplifying assumptions were made.

DEPRECIATION. The declining-balance method of computing depreciation is used because of its favorable tax implications.

TAX RATE. For simplicity, a corporate income tax rate of 50 percent is used throughout.

SEQUENCE OF CALCULATIONS. The procedure is as follows: The model provides for carrying each of four sales-volume estimates for each brand through all calculations that lead to an accumulated present value of expected cash flow. This is done for the four sales-volume estimates for the first brand and for four expected values of sales-volume estimates for the second brand, which are conditional on the identity and the sales volumes of the first brand. (The expected values are computed using inputs called for in Section K of Exhibit 9–1.) The advantage of the procedure is that it yields multiple values of accumulated present value of expected cash flow which, with their probabilities, are indicators of exposure to gain and loss. Had unconditional expected values of the sales-volume estimates for each brand been computed at the outset, only one overall accumulated present value of expected cash flow would have been available.

The program prints out four accumulated present values of expected cash flow for alternatives that call for developing the 100-percent freeze-dried coffee first and eight values for alternatives that focus first developmental efforts on the 20–80 mix. In the latter case, four values are produced under the assumption that the efforts would succeed and four under the assumption that they would fail. The four (or eight) values fully reflect the uncertainty associated with the sales-volume estimates for the first brand, but they do not do so for the estimates for the second brand. Hence, the range of possible consequences for two-brand alternatives is somewhat greater than is indicated by the print-out.

The explanation for the understatement lies in the shortcut represented by the use of the four conditional expected values as certainty equivalents for the second-brand sales volumes (given each of four possible volumes for the first brand). The program could have been written to produce arrays of sixteen (or thirty-two) values of accumulated present value of cash flow for two-brand alternatives. The sixteen values would be based on combinations of each of the four first-brand sales estimates with each of the four sales estimates for the second brand, which are conditional on the attainment of the given first-brand estimate. Similarly, there could have been thirty-two values for two-brand alternatives involving a decision to focus first developmental efforts on the 20–80 mix (separate sets of sixteen for the possibilities of success and failure).

In order to save computation time, the output was restricted to four (or eight) values for each alternative. It was assumed that the shortcut, applied equally to all second brands, would give an adequate comparative measure of risk associated with each two-brand strategy and, in particular, would not affect the final rankings.

STUDY QUESTIONS

ASSIGNMENT 1

1. What are the basic features of the model for prior analysis described in Chapter 9?

2. What do you regard as the model's main advantages? Limitations?

3. Would you recommend any changes in the model? Why or why not?

ASSIGNMENT 2

1. With the aid of information presented in Chapter 6 and 8, prepare the inputs you would like to use if you were to evaluate the alternatives by using the computer program based on the analytical model described in Chapter 9. In this assignment, arrive at the values called for by the following parts of Exhibit 9–1: Sections A (items 1 and 2), I, A (item 3), B, C, D, and E.

2. What are the main assumptions you made in preparing your inputs?

ASSIGNMENT 3

1. Complete your preparation for a prior analysis by arriving at the values called for in Sections F, G, H, I, and K of Exhibit 9–1.

2. What are the main assumptions you made in arriving at the inputs?

ASSIGNMENT 4

1. Using your inputs and the computer program for the model described in Chapter 9, conduct prior analyses of both the one-brand and the two-brand alternatives for the development and marketing of freeze-dried coffee.

2. To what extent do you think the results of your prior analyses should influence the management's decision making?

3. Are there considerations other than those taken into account by the model that you think should influence the choice of the first brand-mix alternatives to be developed? If so, how would you bring them into the decision making?

4. All things considered, what recommendations would you make to the Maxwell House Division management concerning action to be taken to develop and market freeze-dried coffee?

EXHIBIT 9–1

INPUT DATA SHEET FOR PRIOR ANALYSIS OF BRAND-MIX ALTERNATIVES FOR FREEZE-DRIED COFFEE

The inputs required by the model outlined in Chapter 9 are identified below. The exhibit has been prepared to serve as a form on which the reader may enter his own inputs after he has completed Chapter 9 (see Study Questions, Assignments 2 and 3). In Chapter 10, the reader will have an opportunity to compare his inputs and the assumptions on which they are based with those developed by someone else. Important background for the development of inputs was given in Chapters 6 and 8.

A. Enter the following for each brand, assuming it would be the first to be marketed by the Maxwell House Division: (1) sales-volume estimates (SV) in millions of units, which represent the range of expected average going-year (that is, after market introduction) sales prior to the time that outside freeze-dried-coffee competition starts to affect Maxwell House Division sales; (2) probability of occurrence for each sales-volume estimate p(SV); and (3) marketing cost for each level of sales volume (MC|SV) expected after allowing for the effect of outside freeze-dried-coffee competition.

First Brand	SV	P(SV)	MC|SV
Maxwell House	______	______	______
	______	______	______
	______	______	______
	______	______	______
Yuban	______	______	______
	______	______	______
	______	______	______
	______	______	______
Super Yuban	______	______	______
	______	______	______
	______	______	______
	______	______	______
New brand (caffein)	______	______	______
	______	______	______
	______	______	______
	______	______	______
New brand (decaffeinated)	______	______	______
	______	______	______
	______	______	______
	______	______	______

EXHIBIT 9–1 (Continued)

B. Enter for each brand estimates of gross margin per unit (GM) and their probabilities (P).

Brand	GM	P
Maxwell House	______	______
	______	______
	______	______
	______	______
Yuban	______	______
	______	______
	______	______
	______	______
Super Yuban	______	______
	______	______
	______	______
	______	______
New brand (c) [a]	______	______
	______	______
	______	______
	______	______
New brand (d) [a]	______	______
	______	______
	______	______
	______	______

[a] The letters (c) and (d) after "New brand" stand for "caffein" and "decaffeinated," respectively.

C. Enter the following for each brand: (1) introductory marketing costs in millions of dollars (INMKT); (2) total development time in years for first brand (TDEV); (3) time in years required for market testing (TMTEST); (4) plant costs in millions of dollars per million units per year (incremental costs), assuming a 100 percent freeze-dried coffee (PLTCST); and (5) decimal representing the average percentage change in sales volume expected after first year following completion of market introduction (GROWTH).

Brand	INMKT	TDEV	TMTEST	PLTCST	GROWTH
Maxwell House	______	______	______	______	______
Yuban	______	______	______	______	______
Super Yuban	______	______	______	______	______
New brand (c)	______	______	______	______	______
New brand (d)	______	______	______	______	______

EXHIBIT 9–1 (Continued)

D. Enter for each brand the time in years required for market introduction.

Brand	100-Percent Freeze-Dried	(20–80 Mix)
Maxwell House	______	______
Yuban	______	______
Super Yuban	______	______
New brand (c)	______	______
New brand (d)	______	______

E. Enter for each brand the decimals of its first-brand sales volume expected to come from Maxwell House Division ground and soluble coffees.

First Brand	Ground	Soluble
Maxwell House	______ [b]	______ [b]
Yuban	______ [b]	______ [b]
Super Yuban	______	______
New brand (c)	______	______
New brand (c)	______	______

[b] Over base volume

F. Enter for each second brand the decimals of its sales volume expected to come from Maxwell House Division ground, spray-dried, and freeze-dried coffees, given the identity of the first brand.

First Brand	Second Brand	Ground	Spray-Dried	Freeze-Dried
Maxwell House	Yuban	______	______	______
	Super Yuban	______	______	______
	New brand (c)	______	______	______
	New brand (d)	______	______	______
Yuban	Maxwell House	______	______	______
	New brand (c)	______	______	______
	New brand (d)	______	______	______
Super Yuban	Maxwell House	______	______	______
	New brand (c)	______	______	______
	New brand (d)	______	______	______
New brand (c)	Maxwell House	______	______	______
	Yuban	______	______	______
	Super Yuban	______	______	______
	New brand (d)	______	______	______
New brand (d)	Maxwell House	______	______	______
	Yuban	______	______	______
	Super Yuban	______	______	______
	New brand (c)	______	______	______

EXHIBIT 9–1 (Continued)

G. Enter the years in which a freeze-dried coffee marketed by another company might first affect Maxwell House Division coffee sales and the probabilities for those years.

Year	Probability
________	________
________	________
________	________

H. Enter for each brand the decimal of its original (first-brand) sales-volume estimate that is applicable if the brand is second in order, given the identity of the first freeze-dried brand and assuming no outside freeze-dried competition.

First Brand	Second Brand	Decimal
Maxwell House	Yuban	________
	Super Yuban	________
	New brand (c)	________
	New brand (d)	________
Yuban	Maxwell House	________
	New brand (c)	________
	New brand (d)	________
Super Yuban	Maxwell House	________
	New brand (c)	________
	New brand (d)	________
New brand (c)	Maxwell House	________
	Yuban	________
	Super Yuban	________
	New brand (d)	________
New brand (d)	Maxwell House	________
	Yuban	________
	Super Yuban	________
	New brand (c)	________

I. Enter for each brand the decimal of its original (first-brand) sales-volume estimate that is applicable as the average for the period after outside freeze-dried-coffee competition has first affected Maxwell House Division coffee sales.

Brand	Decimal
Maxwell House	________
Yuban	________
Super Yuban	________
New brand (c)	________
New brand (d)	________

EXHIBIT 9–1 (Continued)

J. Enter a figure for each of the following inputs:

1. Depreciation rate for declining-balance method (expressed as decimal) ______
2. Estimate average gross margin for Maxwell House Division ground coffees minus average marketing cost (expressed in dollars per unit) ______
3. Estimated average gross margin for Maxwell House Division soluble coffees minus average marketing cost (expressed in dollars per unit) ______
4. A factor that will be multiplied by net plant to obtain market value of plant at end of last year in period of analysis ______
5. Discount rate (expressed as a decimal) ______
6. Number of years chosen for time period of analysis ______
7. Probability of successfully developing a 20–80 mix ______
8. Rate of return chosen to be high enough to exceed the actual rate of return for all alternatives (needed for rate-of-return computations) ______
9. Code to indicate whether one-brand or two-brand alternatives are to be analyzed (1 = one brand; 0 = two-brand sequence) ______

K. For each of the four levels of estimated sales volume for the second brand, enter its probability, conditional on the attainment by the first brand of each level of its estimated sales volume.

	Sales Estimates for Second Brand			
Sales Volumes for First Brand	Highest	2d Highest	3d Highest	Lowest
Maxwell House		Yuban		
Highest	______	______	______	______
2d highest	______	______	______	______
3d highest	______	______	______	______
Lowest	______	______	______	______
Maxwell House		Super Yuban		
Highest	______	______	______	______
2d highest	______	______	______	______
3d highest	______	______	______	______
Lowest	______	______	______	______
Maxwell House		New brand (c)		
Highest	______	______	______	______
2d highest	______	______	______	______
3d highest	______	______	______	______
Lowest	______	______	______	______
Maxwell House		New brand (d)		
Highest	______	______	______	______
2d highest	______	______	______	______
3d highest	______	______	______	______
Lowest	______	______	______	______

EXHIBIT 9–1 (Continued)

Sales Volumes for First Brand	Sales Estimates for Second Brand: Highest	2d Highest	3d Highest	Lowest
Yuban		Maxwell House		
Highest	______	______	______	______
2d highest	______	______	______	______
3d highest	______	______	______	______
Lowest	______	______	______	______
Yuban		New brand (c)		
Highest	______	______	______	______
2d highest	______	______	______	______
3d highest	______	______	______	______
Lowest	______	______	______	______
Yuban		New brand (d)		
Highest	______	______	______	______
2d highest	______	______	______	______
3d highest	______	______	______	______
Lowest	______	______	______	______
Super Yuban		Maxwell House		
Highest	______	______	______	______
2d highest	______	______	______	______
3d highest	______	______	______	______
Lowest	______	______	______	______
Super Yuban		New brand (c)		
Highest	______	______	______	______
2d highest	______	______	______	______
3d highest	______	______	______	______
Lowest	______	______	______	______
Super Yuban		New brand (d)		
Highest	______	______	______	______
2d highest	______	______	______	______
3d highest	______	______	______	______
Lowest	______	______	______	______
New brand (c)		Maxwell House		
Highest	______	______	______	______
2d highest	______	______	______	______
3d highest	______	______	______	______
Lowest	______	______	______	______
New brand (c)		Yuban		
Highest	______	______	______	______
2d highest	______	______	______	______
3d highest	______	______	______	______
Lowest	______	______	______	______

EXHIBIT 9–1 (Concluded)

Sales Volumes for First Brand	Sales Estimates for Second Brand: Highest	2d Highest	3d Highest	Lowest
New brand (c)		Super Yuban		
Highest	______	______	______	______
2d highest	______	______	______	______
3d highest	______	______	______	______
Lowest	______	______	______	______
New brand (c)		New brand (d)		
Highest	______	______	______	______
2d highest	______	______	______	______
3d highest	______	______	______	______
Lowest	______	______	______	______
New brand (d)		Maxwell House		
Highest	______	______	______	______
2d highest	______	______	______	______
3d highest	______	______	______	______
Lowest	______	______	______	______
New brand (d)		Yuban		
Highest	______	______	______	______
2d highest	______	______	______	______
3d highest	______	______	______	______
Lowest	______	______	______	______
New brand (d)		Super Yuban		
Highest	______	______	______	______
2d highest	______	______	______	______
3d highest	______	______	______	______
Lowest	______	______	______	______
New brand (d)		New brand (c)		
Highest	______	______	______	______
2d highest	______	______	______	______
3d highest	______	______	______	______
Lowest	______	______	______	______

EXHIBIT 9–2

INSTRUCTIONS FOR PREPARING INPUT DATA CARDS FOR PRIOR ANALYSIS OF FREEZE-DRIED COFFEE ALTERNATIVES

The computer program was written to use ten fields of data per card, each field consisting of five columns. Punch only one set of data in each field. Begin with the field at the left-hand margin and move across to the right, using as many fields as you have sets of data. Each number that represents a quantity rather than a code should include a decimal point. All numbers should be right-justified.

Punch the appropriate brand code in the first field of all cards that relate to a particular brand. Brand data cards should be run in the order of the brand code numbers. Order of cards that relate to the same brand will be specified later in the instructions.

A. FIRST TWENTY CARDS. Each brand will have four cards. Brand 1 (Maxwell House) will be represented by the first four cards; brand 2 (Yuban) by the second four cards, and so forth. For each brand, punch four cards so that each contains the following:

1. One of four sales-volume estimates in millions of units per year that you select to represent the range of expected average going-year (that is, after market introduction) sales volumes prior to the time that outside freeze-dried-coffee competition starts to affect Maxwell House Division sales

2. The probability of occurrence for each sales-volume estimate

3. Marketing cost for each level of sales volume expected after allowing for the effect of outside freeze-dried-coffee competition

The four cards for each brand should appear in the following order:

First card: SV_1 (highest), $P(SV_1)$, MC_1
Second card: SV_2 (2d highest), $P(SV_2)$, MC_2
Third card: SV_3 (3d highest), $P(SV_3)$, MC_3
Fourth card: SV_4 (lowest), $P(SV_4)$, MC_4

The sales-volume estimates are to be made under the following assumptions:

That the brand would be the first freeze-dried coffee on the market

That the brand's market introduction would start earlier than the market introduction of a freeze-dried coffee marketed by a competitor by the following amounts, provided Maxwell House's initial product-development attempt (whether for a 100-percent freeze-dried coffee or a 20–80 mix) is successful: for Maxwell House, 3½ years; for Yuban, Super Yuban, and new brand (decaffeinated), 4 years; for new brand (caffein), 4½ years

Assumptions you regard as appropriate about the relative retail prices of freeze-dried, spray-dried, and ground coffees

Your sales estimates may fully allow for growth in volume expected over the entire period of analysis. If you prefer, however, the estimates may represent going-year volumes attained in the first year after market introduction has been

EXHIBIT 9–2 (Continued)

completed. You can then allow for any growth expected in the remaining years in the period of analysis by means of rate-of-growth inputs provided for in Section C, item 5.

Marketing costs are to be conditional on sales volume expected after allowing for the effect of outside freeze-dried-coffee competition. Therefore, you will want to make the marketing-cost estimates after completing Section I of the input data.

B. NEXT FIFTEEN CARDS. For each brand, prepare three cards so that each contains an estimate of gross margin and its probability of occurrence. The cards should be in the following order:

First card: GM_1 (highest), $P(GM_1)$
Second card: GM_2 (middle), $P(GM_2)$
Third card: GM_3 (lowest), $P(GM_3)$

C. NEXT FIVE CARDS. For each brand, enter a point estimate for each of the following items:

1. Introductory marketing costs in millions of dollars.
2. Total development time in years for the brand, assuming that it would be the first brand developed.
3. Time in years required for market testing.
4. Plant costs in millions of dollars per million units per year (use only incremental costs), assuming a 100 percent freeze-dried coffee.
5. Decimal representing the average annual percentage change in sales volume expected after first year following completion of market introduction. (If you fully allowed for growth in volume expected over the entire period of analysis in making your going-year sales-volume estimates called for in Section A, punch in .0.)

One card for each brand should contain all five pieces of information in the first five fields in the order listed above.

D. NEXT FIVE CARDS. Punch the following information on one card for each brand:

1. Time in years required for market introduction, given a 100-percent freeze-dried coffee
2. Time in years required for market introduction, given a 20–80 mix

E. NEXT FIVE CARDS. Punch the following estimates on one card for each of the five brands. Each estimate is to use the assumption that the brand would be the first freeze-dried coffee on the market.

1. The decimal representing the percent of the brand's sales volume expected to come at the expense of Maxwell House Division ground coffees
2. The decimal representing the percent of the brand's sales volume expected to come at the expense of other Maxwell House Division soluble coffees

EXHIBIT 9–2 (Continued)

F. NEXT EIGHTEEN CARDS. Punch one card to contain the following information in the order listed for each two-brand sequence to be considered (in this case, brand cards do not have to appear in a given order):

1. Code for first brand
2. Code for second brand
3. The decimal representing the percent of second brand's sales volume expected to come at the expense of Maxwell House Division ground coffees
4. The decimal representing the percent of second brand's sales volume expected to come at expense of Maxwell House Division spray-dried soluble coffees
5. The decimal representing the percent of second brand's sales volume expected to come at expense of other Maxwell House Division freeze-dried soluble coffees

G. NEXT CARD. Punch one card to identify the years in which you think a freeze-dried coffee marketed by another company might first affect Maxwell House Division coffee sales and the probabilities for those years. (Assume competitive entry as of January 1 of each year designated.)

1. Earliest year possible
2. Probability for earliest year possible
3. Median year possible
4. Probability for median year
5. Latest year possible
6. Probability for latest year

H. NEXT EIGHTEEN CARDS. Punch the following information on one card for each two-brand sequence to be analyzed. (In this instance, no special order of cards is required.)

1. Code for first brand
2. Code for second brand
3. Decimal representing the percent of the original (first-brand) sales-volume estimate for the brand that is to be second in order which you think still applies, given the identity of the first freeze-dried-coffee brand and assuming no outside freeze-dried-coffee competition

I. NEXT CARD. For each brand, punch one card to show the percent of the original (first-brand) sales estimate you think would represent the average sales volume for the brand after outside freeze-dried competition has first affected Maxwell House Division sales. In arriving at the estimates, you are to make assumptions that you regard as appropriate about the nature, extent, and timing of outside freeze-dried-coffee competition.

EXHIBIT 9–2 (Concluded)

J. NEXT CARD. This card should contain the following nine items of information in the order given:

1. Depreciation rate for declining-balance method
2. Estimated average gross margin for Maxwell House Division ground coffees minus average marketing cost (expressed in dollars per unit)
3. Estimated average gross margin for Maxwell House Division soluble coffees minus average marketing cost (expressed in dollars per unit)
4. A factor that will be multiplied by net plant to obtain market value of plant at end of last year in period of analysis
5. Discount rate (expressed as a decimal)
6. Number of years chosen for the time period of the analysis. (You are limited to a maximum of 15 years.)
7. Probability of successfully developing a 20–80 mix. (The program assumes that a 100-percent freeze-dried coffee can be successfully developed for certain.)
8. A rate of return chosen to be high enough to exceed the actual rate of return for all alternatives being considered
9. Code to indicate whether one-brand or two-brand alternatives are to be analyzed. (If only one brand, punch 1; if two-brand sequence, punch 0.)

K. LAST SEVENTY-TWO CARDS. Assign conditional probabilities for each of your second-brand sales-volume estimates, assuming, in turn, that each sales volume estimated for each first-brand alternative was realized. In other words, you need to fill out a matrix like the one given below for each of the eighteen two-brand sequences under consideration.

PROBABILITIES OF SALES ESTIMATES FOR SECOND BRAND, CONDITIONAL ON SALES VOLUMES OF FIRST BRAND

Sales Volume for First Brand	Sales Estimates for Second Brand			
	Highest	2d Highest	3d Highest	Lowest
Highest	______	______	______	______
2d highest	______	______	______	______
3d highest	______	______	______	______
Lowest	______	______	______	______

Punch four cards, one for each row of each matrix. The first field should contain the code for the first brand; the second field should contain the code for the second brand; and the third, fourth, fifth, and sixth fields should contain the conditional probabilities for the highest, second highest, third highest, and lowest sales estimate, respectively, for the second brand. The four cards for a given two-brand sequence should appear in the following order of sales-volume estimates for the first brand: highest, second highest, third highest, and lowest.

CHAPTER 10

PRIOR ANALYSIS OUTPUT AND MANAGEMENT DECISION

The output of prior analyses of the one-brand and two-brand alternatives for the development and marketing of freeze-dried coffee are presented in this chapter. The analyses were made using the analytical model described in Chapter 9 and the input data that appear in Exhibit 10–1. Assumptions of the input data and the model itself will be reviewed prior to considering the extent and nature of the influence that the results should have on management decision. Attention will then be given to several ways in which the decision model might be employed by management.

PRIOR ANALYSIS OUTPUT FOR TWO-BRAND ALTERNATIVES

Accumulated present values of cash flow for two-brand alternatives, based on ten- and fifteen-year time periods, appear in Exhibit 10–2. They lead to the following observations:

1. The profit potential of the 20–80 mix is substantially greater than that of 100-percent freeze-dried coffee even after allowing for the former's much lower probability of developmental success. The finding importantly reflects the assumptions of lower costs for the mix and independence of sales volume and choice of mix.

2. The outlook for increased Division profits improves dramatically between ten and fifteen years hence, reflecting the anticipated growing sales volumes and lower costs. None of the two-brand alternatives would

lead to improved profits within ten years if 100-percent freeze-dried coffee were used. The same can be said for all but eight of the alternatives if the initial development decision is for the 20–80 mix.

3. The rank orders of the leading two-brand alternatives are approximately the same whether based on ten- or fifteen-year periods of analysis. Attention will be focused on fifteen-year output, which more fully reflects anticipated sales growth and effects of competition.

4. On the basis of accumulated present value of expected cash flow, a new brand (with caffein) should be the first and a new brand (decaffeinated) should be the second to be developed and marketed. The value for that two-brand sequence for the fifteen-year time period and the 20–80 mix is $42.84 million. It compares with $41.04 million for new brand (decaffeinated)–new brand (caffein), $36.31 million for new brand (decaffeinated)–Super Yuban, and $32.19 million for Super Yuban–new brand (decaffeinated).

Rates of return and profitability-index values, based on the fifteen-year period of analysis, are given in Exhibit 10–3. They prompt the following observations:

1. The rank order of the alternatives is approximately the same whether based on rate of return or on the profitability index. As it turned out, single rates of return could be calculated for all alternatives, so that the need for the profitability index was less than it otherwise would have been.

2. The four leading alternatives have approximately the same rates of return (ranging from 24.5 to 25.8) and profitability-index values (ranging from 1.588 to 1.618).

3. The top four alternatives as identified by each of the three measures (accumulated present value of expected cash flow, rate of return, and profitability index) are the same, but their ranks differ (see Exhibit 10–4). Super Yuban–new brand (decaffeinated) placed fourth on accumulated present value of expected cash flow ($32.19 million) but first on rate of return (25.8 percent). New brand (caffein)–new brand (decaffeinated) placed first on accumulated present value of expected cash flow ($42.84 million) but fourth on rate of return (24.5 percent).

Probability distributions indicating exposure to gain and loss of the four top alternatives as identified by the measures mentioned above are given in Exhibit 10–5. Assuming an initial development decision for the 20–80 mix, new brand (caffein)–new brand (decaffeinated) offers exposure to the largest gain ($171 million); whereas Super Yuban–new brand (decaffeinated) is fourth in this respect ($116 million). Super Yuban–new brand (decaffeinated), however, has the highest minimum expectation ($11 million), whereas new brand (caffein)–new brand (decaffeinated) is second ($9 million). These figures are based on a fifteen-year period of analysis.

Using a ten-year time period and the 20–80 mix, new brand (caffein)–new brand (decaffeinated) offers exposure to the largest gain ($86 million) and the largest loss ($8 million). Super Yuban–new brand (decaffeinated) offers exposure to the smallest loss ($5 million) and the smallest gain ($53 million) represented by the top four alternatives. Choice of alternative, then, will depend in part on the decision maker's attitude toward risk.

PRIOR ANALYSIS OUTPUT FOR ONE-BRAND ALTERNATIVES

Selected results of prior analysis of one-brand alternatives based on a fifteen-year time period are shown in Exhibit 10-8. They include values of accumulated present value of expected cash flow, rate of return on investment, and profitability index. New brand (decaffeinated) ranks first on all counts, regardless of choice of mix. If the first development effort were for the 20–80 mix, new brand (caffein) would rank second, ahead of Super Yuban on accumulated present value of expected cash flow, but third behind Super Yuban on rate of return and profitability index. If the initial development decision were for 100-percent freeze-dried coffee, new brand (caffein) would rank second ahead of Super Yuban on all three measures.

A prior analysis using the same inputs but a ten-year time period showed that new brand (decaffeinated) ranked first, new brand (caffein) ranked second, and Super Yuban ranked third on all three measures, assuming an initial development decision for the 20–80 mix. If first development efforts were for 100-percent freeze-dried coffee, new brand (decaffeinated) ranked first on all three measures. Super Yuban ranked second, ahead of new brand (caffein) on accumulated present value of expected cash flow, but third behind new brand (caffein) on rate of return and profitability index.

Perhaps the most important observation to be made here is that the prior analysis of one-brand alternatives seemed to point to a different first-brand choice than the prior analysis of two-brand sequences did.

HOW SHOULD PRIOR ANALYSIS OUTPUT INFLUENCE MANAGEMENT DECISION?

The answer, of course, depends on the extent to which management can accept the results as being indicative of the relative merits of the alternatives. That, in turn, depends on the following considerations:

1. The extent to which management is satisfied with the judgments reflected in the input data
2. The extent to which management is satisfied with the assumptions incorporated in the analytical model

3. The extent to which management feels that important decision criteria have been satisfied

These considerations are discussed in the sections that follow.

ASSUMPTIONS OF THE INPUT DATA

Most of the assumptions represented by the inputs (see Exhibit 10–1) are readily apparent. Some are not, however, and they will be discussed here.

SELLING PRICES. The selling prices assumed in estimating sales and gross margins can be seen by comparing the gross-margin inputs listed in Exhibit 10–1 with information given in Exhibit 8–2. The price premiums represented by the assumptions were in the following ranges: Maxwell House, 5 to 20 percent and new brand (caffein), 7 to 25 percent more than the price charged for Maxwell House spray-dried coffee; Yuban, 5 to 20 percent and Super Yuban, 5 to 25 percent more than the price charged for Yuban spray-dried coffee; and new brand (decaffeinated), 10 to 27 percent more than the price charged for Sanka spray-dried coffee.

SALES-VOLUME ESTIMATES. In estimating sales volume for the first freeze-dried brand, it was assumed that its market introduction would take place earlier than that of a competing freeze-dried coffee by the following amounts, provided that the Maxwell House Division's initial development attempt (whether for 100-percent freeze-dried coffee or a 20–80 mix) succeeded: Maxwell House, 3½ years; Yuban, Super Yuban, and new brand (decaffeinated), 4 years; and new brand (caffein), 4½ years. It also was assumed that outside freeze-dried-coffee competition would come primarily in the form of brands directly competitive with Maxwell House and new brand (caffein).

It was assumed that new sales-volume estimates would not be needed for the brand alternatives in the event of failure of an initial developmental effort on the 20–80 mix. It was expected that such failure would result in a two-year delay in starting work on a 100-percent freeze-dried product for which developmental success was assumed. Under the two-year delay, Yuban, Super Yuban, new brand (caffein), and new brand (decaffeinated) still would have completed market introduction before outside competition began to affect Maxwell House Division sales even if such effect came in the earliest year in which it was expected. The same brands, if second in order, would have completed market introduction before outside freeze-dried competition affected Maxwell House Division sales if the effect came in the medium expected year. If it came in the earliest expected year, market introduction of the four brands would be from 60 to 100 percent complete. The four brands would be similarly situated in this regard, so they would be affected in about the same proportions. Therefore, new sales-volume estimates did not appear to be needed for the purposes of a relative analysis.

In the event of a two-year delay occasioned by failure of an initial product-development attempt focused on the 20–80 mix, it was estimated that Maxwell House brand freeze-dried coffee, if first in order, would have completed 63, 88, and 100 percent of its market introduction before the effect of outside freeze-dried competion would be felt if the latter came in the earliest, medium, and latest expected years, respectively. If Maxwell House were to be second in order, its market introduction would be 44 to 56 percent, 69 to 81 percent, or 94 to 100 percent complete before the effect of outside competition would be first felt, assuming that the latter started in the earliest, medium, and latest expected years, respectively. If Maxwell House were to emerge as a serious contender among the five brands under consideration, new sales estimates would be needed that allowed for the effect of a two-year delay. Maxwell House, however, was not expected to rank that high.

GROWTH. The going-year sales estimates were made by visualizing a redistribution of the volume represented by the total coffee market that would occur immediately after introduction of the first freeze-dried brand. The estimates did not allow for growth in total coffee demand that would take place over the time period of the analysis or any expectations that the rates of increase for different freeze-dried brands would differ among themselves and from the rates for all coffee and the freeze-dried-coffee category. To make such allowances, the following rates of growth were applied after the first year following completion of market introduction: Maxwell House, 4 percent; Yuban and Super Yuban, 6 percent; new brand (caffein) and new brand (decaffeinated), 10 percent.

The inputs used in the analysis reported in this chapter represent judgments with which others may or may not agree. If the decision maker prefers different values, he should consider whether the changes they represent are likely to affect the final decisions on whether the first developmental efforts should be on the 20–80 mix or the 100-percent freeze-dried product and which two-brand sequence should be adopted. Those who have access to a computer for work on this problem can run their own inputs and compare the results with those reported here. For purposes of illustration, the inputs in Exhibit 10–1 will be accepted at this point so that we may continue by giving further consideration to those aspects of the model itself that bear on the question of the acceptability of the output.

ASSUMPTIONS OF THE MODEL

As in the case of the input data, management has the final responsibility for determining the acceptability of the assumptions of the model. Substantial disagreement, of course, could lead to rejection of the output, at least until the model was revised. For our example, we shall assume that most of the model's assumptions are satisfactory for one or more reasons: they seem to be in accord with observed reality; they reflect requisite judgments of persons in whom the decision maker has confidence; and, within reasonable limits, changes in the assumptions are unlikely to affect the relative values by which the alternatives are to be compared. Other assumptions may be tentatively accepted, subject to review upon receipt of the output. Relative to the evaluation, additional comments on a few of the model's assumptions that were identified in Chapter 9 are appropriate here.

RELEVANCY OF OUTPUT. This assumption warrants further attention because of its basic character. The model was designed to produce four relative measures: (1) change in the financial position of the Maxwell House Division expressed as accumulated net present value of expected cash flow, (2) expected internal rate of return, (3) profitability index, and (4) exposure to gain and loss indicated by values representing the range of profit consequences and their probabilities.

The measures represent decision criteria commonly regarded as fundamental. The first three are based on different mathematical formulas for arriving at a single expression of value. Although they have the objective of evaluation in common, they are not equivalent. Hence, they may lead to different rankings of alternative investment proposals. All three measures have their advocates, and there is no one right answer concerning which should be used. All three were included here for purposes of illustration and to emphasize the need for clear specification of decision criteria. There would be no need, of course, to design the model to produce all of the first three measures listed if the decision maker had determined that he should use only one or two of them in combination with information on exposure to gain and loss.

The measures of accumulated net present value of cash flow and the profitability index involve discounting cash flows to their present values, thereby taking into account both the size and timing of the cash inflows and outflows.

Computation of accumulated net present value of cash flow requires a decision on a rate of return to be used in discounting. In this case, the rate used was the estimated cost of capital to the firm. The choice of discount rate is not inconsequential. The application of different rates to different time schedules of cash flows can result in different rankings of the alternatives. The net present value of cash flow is an absolute measure that describes the profit potential of an investment, given the rate of interest used in the discounting.

The internal-rate-of-return method involves solving for a rate of return, namely, that rate which makes the net present value of cash flow zero.

The latter is a measure of the investment efficiency that can be used in comparing alternatives.

Both net present value of cash flow and internal rate of return are correct measures for determining the worth of a given investment opportunity. However, they may produce different rankings of mutually exclusive alternatives because of differences in assumptions about the rate at which funds produced by the project are reinvested. The present-value method implies reinvestment at a rate equivalent to the required rate of return used as the discount rate. The internal-rate-of-return method implies that funds are reinvested at the internal rate of return over the remaining life of the project. The decision maker must decide which assumption is more appropriate.

In computing cash flow, depreciation is not subtracted from receipts. Expected true depreciation, however, is taken into account by the inclusion of an estimated salvage value of the investment at the end of the time period. Distinguishing true income from capital recovery is important for tax and financial reporting but is not necessary for computing the annual rate of return promised by a project over its economic life.

A computational problem that may be encountered with the internal-rate-of-return method is that it usually is not possible to compute a unique internal rate of return if cash flows are irregular (that is, if they change from positive to negative to positive over time). A correct solution in such a situation requires approaching the problem differently.[2] This computational disadvantage is not shared by accumulated net present value of expected cash flow or the profitability index. The latter is an investment efficiency measure consisting simply of the ratio of the present value of net cash inflows from operations to the present value of investment outlays.

Several other methods that have been used to evaluate new investments were seen as inappropriate for the problem at hand. Among them is the widely used payback period, which is a time rather than a rate concept. The payback period is useful when liquidity as opposed to profitability is an important short-run consideration.

The three measures of profitability used (accumulated present value of cash flow, internal rate of return, and profitability index) are single net expressions of outcome. They do not directly indicate exposure to gain and loss. Two alternatives might vary considerably in that respect while appearing to be about equal on other counts. Hence, the provision was made for gauging such exposure in terms of accumulated present value of expected cash flow so that it might be considered in the context of the decision maker's attitude toward risk.

[2] See Ezra Solomon, *The Theory of Financial Management* (New York: Columbia University Press, 1963), pp. 128–131; or James C. Van Horne, *Financial Management and Policy* (Englewood Cliffs, N.J.: Prentice-Hall, Inc., 1968), pp. 51–53.

Although the four measures may be accepted as being significant, the model does not provide for weighting them. That task is left to the decision maker. Management may regard as important other decision criteria that are not served by the model. Share-of-market objectives and a desire to enhance a reputation for innovative leadership are examples. It is up to the decision maker to see that appropriate recognition is given to all criteria he regards as relevant.

INVESTMENT. The procedure for handling investment in a given year as the cost of the additional plant needed to produce the next year's output of freeze-dried coffee is one that should be reexamined with the output in hand. It may or may not be satisfactory for the purpose of the relative analysis. It could lead to overbuilding of plant if sales in the earlier years were expected to be substantially greater than those later on because of growing competition. The amount of such overbuilding could vary by alternative. Hence, the yearly sales and investment figures should be checked to determine whether use of the investment assumption unduly handicaps any of the more promising alternatives.

The print-out for new brand (caffein)–new brand (decaffeinated) based on the 20–80 mix is given in Exhibit 10–6. It shows that sales in millions of units for new brand (caffein) increase to 10.58 in 1967, decline to 6.42 in 1970, and rise again to 11.52 in 1976. The procedure of annually adding sufficient plant to produce next year's sales, therefore, results in unused capacity in eight years, assuming that separate plants are needed for the two types of freeze-dried coffee that the brands represent. The maximum excess capacity in a year is 4.16 million units, and the figure ranges from about 2 to 3.5 million units in four other years.

Although some unused plant capacity probably is unavoidable under an optimum construction schedule, prior analysis computations give the new brand (caffein)–new brand (decaffeinated) alternative considerably more excess capacity than the other leading two-brand sequences (see Exhibit 10–7). It appears, therefore, that it has suffered disproportionately. It probably would benefit substantially more than the other top three contenders from a more realistic and economical schedule of investment. Such a modification might result in its ranking first instead of fourth on rate of return, especially since the rates computed for the four highest-ranking alternatives were not far apart.

New brand (decaffeinated) involved no unused plant capacity under the model's handling of investment, but the other brands did.

MIX. The model uses the judgment that if the first developmental effort were focused on the 20–80 mix and it failed, successful completion of such an attempt at a later time could not take place much before the expiration of the time period of the analysis. If the first try succeeded, the 20–80 mix would be used because of its cost advantages. Hence the provision that both the first and second brands would employ either 100-percent freeze-dried coffee or the 20–80 mix, whichever was developed first. If development of the 20–80 mix subsequent to the development of 100-percent freeze-dried coffee was seen as a possibility before the end of the time period, this factor should be taken into account. Plant requirements for the mix were only one-third of those of the 100-percent freeze-dried product. Because of their differences in expected sales volume, some alternatives would involve greater risk than others of building plant that would no longer be needed by the two brands under consideration once the 20–80 mix was available.

SHOULD THE MODEL BE ELABORATED?

Although the computer program runs sixteen print-out pages in length, the model is reasonably simple in its logic. It became more elaborate as work on it continued over time; but, of course, it remains a simplification of the real world. The question is whether it is too much of a simplification for the decision task at hand.

The model could be changed to reduce dependency on some of the assumptions, as has been indicated already. Investment, for example, could be handled in a manner reflecting consideration of optimum plant size, the amount of capacity that should be built at any one time, construction time, and related factors.

Management already had made the decision to go ahead with the freeze-dried-coffee project and expected that it would be successful. Success, however, was not a certainty. The model could be elaborated to allow for the probability and the consequences of later abandonment. Whether such a modification is warranted for the purpose of choosing the mix and the first and second brands is not clear on the basis of readily available information. The 100-percent freeze-dried product has higher capital requirements than the 20–80 mix, but it also has a higher probability (certainty was assumed) of developmental success and, therefore, early market entry. The brand alternatives differ in expected sales volumes and, therefore, in total capital requirements. The extent to which they also may vary on risk of loss in case of market failure depends on the size and timing of the plants that would be constructed.

The model also might be made more complete in ways that would call for somewhat different inputs from the decision maker. For example, a provision could be made for inputs of retail prices and cost of goods sold so that the computer would calculate gross margin per unit, which now must be entered as an input. The model now works with sales volumes, costs, and gross margins, which are really weighted averages of the several sizes of packages. The model could be modified to receive estimates of these variables and to compute the weighted averages. Whether better inputs would be produced if more detailed forecasting were required would depend upon the decision maker (with his assistants) and the information he had to go on. Working with estimates in terms of averages per ounce is simpler. A high degree of accuracy is improbable for forecasts for up to ten to fifteen years into the uncertain future, and it is not requisite to choosing among the alternatives.

The model could be made to receive annual sales-volume estimates of competitive freeze-dried-coffee volume by price level and type (caffein versus decaffeinated) and judgments concerning how the sales of Maxwell House Division freeze-dried brands would be affected by the different kinds of competition. As the model now stands, the decision maker must enter inputs that represent his judgment about the average effect on Maxwell House Division freeze-dried-coffee sales of competitive freeze-dried coffees for the entire period in which the latter are expected to be on the market.

The model could be elaborated into a simulation in which the computer would randomly select a value from the probability distribution for each variable to use in a given run of the computer program. With such a model, a number of runs could be made to establish the probable range of consequences that could be expected from the different courses of action.

The model could also be changed to accommodate inputs of all cost information so that the final measures would be absolute rather than relative. Although the added information may be of value for other reasons, a relative analysis is adequate for choosing among the alternatives, given the assumption that management has decided to go ahead.

In considering alterations in the model, its purpose should be kept clearly in mind. The immediate purpose is that of providing a practical basis for choosing the best two-brand course of action. For this objective, then, changes should be advocated because someone believes there is a reasonable probability that they would lead to a different ranking of alternatives than would be obtained otherwise. Admittedly, the importance of a change may be difficult to judge in advance of making it. Executives may differ in their opinions depending on their backgrounds, customary ways of thinking, and access to information on which to base inputs. The model described in Chapter 9 represents someone's judgment that it would be adequate. Actually, one could argue that it is more elaborate than is necessary for the decision task at hand though not for other uses to be mentioned in the next section of this chapter.

A useful test of a formal analysis consists of asking whether the principal results can be explained by informal reasoning. If they can not, the analysis should be examined closely. Faulty structuring or technical errors may be discovered. If not, the search for them is likely to lead to improved insight, modifications of the earlier intuitive judgment, and greater confidence in the formal analysis output.

In our example, the results seem to make sense, given key assumptions and, we should add, an understanding of the problem that has been gained from structuring it for analysis. At first one might be surprised that the 20–80 mix, with its substantially lower probability of developmental success, did so much better than 100-percent freeze-dried coffee. The result can be understood, however, in terms of its marked cost advantages and the judgments that the final product, regardless of mix, would be of the same quality and, therefore, enjoy the same consumer demand; that if the first try were for the 20–80 mix and it failed, only a two-year delay would result in marketing a successful 100-percent freeze-dried coffee; and that outside freeze-dried-coffee competition would not appear sooner than expected. Similarly, it is not surprising that the two new brands were found to offer the greatest incremental sales and profits given the assumptions about their sales volumes, their cannibalism of sales of other Maxwell House Division coffees, and the nature and timing of outside freeze-dried-coffee competition.

If the decision maker takes exception to key assumptions such as these, his willingness to accept the output will depend on how important he regards his points of criticism to be to the final results. Serious disagreement could lead to rejection and, perhaps, changes in assumptions and a new analysis. In other cases, use of the output along with "eyeball adjustments" could be satisfactory for making the required decision on a course of action.

COMMENTS ON THE OUTPUT AND USES OF THE DECISION MODEL

The model described in Chapter 9 provides a means of systematically taking into account the effects of many factors that vary over time, an important advantage in a complex decision problem. The print-out gives management a year-by-year picture of the net results of these factors and the values of some of them. The model requires a good deal of specific thinking, more than typically takes place in less-formal analyses. The greater rigor frequently may be regarded as advantageous for complex decision problems when a great deal is at stake. As in any type of analysis, the quality of the assumptions used is critical to the value of the results.

In this case, the model's output did not provide an easy, ready-made decision. Additional thought and judgments were required. Different two-brand alternatives ranked first depending on which measure or combination of measures was employed. In addition to the responsibility for evaluating (and participating in the development of) the model and the input data, management faced the task of weighting the different measures of output and considering relevant factors that may not have been fully recognized by the model itself. Any factor that could make for a substantial difference among alternatives is of concern. Small differences in expected outcomes of the alternatives would suggest that it makes little difference which one is chosen, that they are about equal bets.

Once a model has been developed, the computer makes it possible to execute elaborate analyses in a short space of time and at a reasonable cost. The analyses reported in this chapter were made on an IBM 360/67. Computer time for one run for two-brand alternatives on both ten- and fifteen-year time periods was about seventy-seven seconds, and the cost was about $15. Similar runs for one-brand alternatives required twenty-eight seconds of computer time and cost about $6.70.

A much greater cost is represented by the time and the dollars required for conceptualizing the model and preparing the computer program. A rough estimate can be made on the basis of the 900 instructions contained in the program. Assuming that an average programmer can handle five instructions a day (planning and writing them in computer language and preparing flow diagrams and input data instructions), the programming represents 180 man-days of work exclusive of additional time required for conceptualizing the overall approach. This means that even if several men were to work together, three to four months would be required for the development of the program and its documentation. In this case, such an expenditure was justified by the importance of the decision to be made.

In addition to serving the immediate need for a choice among the mix and brand alternatives, the model has other potential uses. It can be used for sensitivity analyses in which the values for factors are changed over a series of runs. For example, in the analysis reported in this chapter, separate runs were made for two different time periods (ten and fifteen years). Similarly, values for factors such as sales volume and gross margin could be systematically varied to determine how much change could take place in them before the final choice of a course of action would be affected. Factors to which the output was found to be especially sensitive then can be given special attention to make sure that their inputs are the best that can be made. The feasibility of subjecting such variables to research can be investigated.

The possibility also exists for using the model to evaluate different marketing strategies and programs by simulation. The programs, for example, might call for different prices or different promotional expenditures and plans for influencing coffee drinkers to try the new product. The use of the model for this purpose, of course, requires a basis for relating the different marketing actions to sales outcomes.

Longer-run potential benefits of the use of decision models include a beneficial effect on executive thought processes, a highlighting of needs for specific information on which to base inputs, and an improved understanding of the nature of relationships among factors important to one's business. The latter may result over time as a product of the specific thinking and systematic inquiries and observations that use of such models can encourage.

STUDY QUESTIONS

1. On the basis of available information, which brand-mix alternative for freeze-dried coffee do you think the Maxwell House Division should attempt to develop first? Which should be developed second? Why?

2. Would you favor the required use of the Bayesian approach on major choice problems if you were general manager of the Maxwell House Division? If you were a product manager who reported to the marketing manager? Why or why not?

EXHIBIT 10–1

INPUT DATA USED IN PRIOR ANALYSIS OF BRAND-MIX ALTERNATIVES FOR FREEZE-DRIED COFFEE[a]

A. Sales-volume estimates (SV) for first brand in millions of units selected to represent the range of expected average going-year (that is, after market introduction) sales prior to the time that outside freeze-dried-coffee competition starts to affect Maxwell House Division sales; probability of occurrence for each sales-volume estimate [P(SV)]; and marketing cost for each level of sales volume expected after allowing for effect of outside freeze-dried-coffee competition (MC|SV):

First Brand	SV	P(SV)	MC|SV
Maxwell House	30.5	.20	1.40
	26.5	.50	1.55
	21.2	.20	1.70
	17.5	.10	1.80
Yuban	8.8	.10	1.65
	7.0	.30	1.70
	5.0	.40	1.80
	3.0	.20	2.00
Super Yuban	9.3	.10	1.50
	7.0	.10	1.55
	5.0	.30	1.70
	3.0	.50	1.80
New brand (c)[b]	18.0	.10	1.55
	14.0	.20	1.60
	10.0	.40	1.70
	7.0	.30	1.90
New brand (d)[b]	11.5	.10	1.55
	10.5	.40	1.60
	9.5	.40	1.75
	8.5	.10	1.95

[a] The input data in this exhibit are presented for illustrative purposes. They represent judgments with which you may or may not agree. The reader is encouraged to develop his own inputs.

[b] The letters (c) and (d) after "New brand" stand for "caffein" and "decaffeinated," respectively.

EXHIBIT 10–1 (Continued)

B. Estimated gross margins per unit (GM) and their probabilities (P):

Brand	GM	P
Maxwell House	$3.60	.25
	3.10	.50
	2.70	.25
Yuban	3.70	.25
	3.10	.40
	2.65	.35
Super Yuban	4.10	.25
	3.40	.40
	2.90	.35
New brand (c)	3.95	.25
	3.45	.40
	3.10	.35
New brand (d)	4.40	.25
	3.75	.50
	3.30	.25

C. Introductory marketing costs in millions of dollars (INMKT); total development time in years for first brand (TDEV); time in years required for market testing (TMTEST); plant costs in millions of dollars per million units per year (incremental costs), assuming a 100-percent freeze-dried coffee (PLTCST); decimal representing the average percentage change in sales volume expected after first year following completion of market introduction (GROWTH):

Brand	INMKT	TDEV	TMTEST	PLTCST	GROWTH
Maxwell House	21.	2.5	1.	5.0	0.04
Yuban	9.	2.0	1.	5.0	0.06
Super Yuban	6.	2.0	1.	5.0	0.06
New brand (c)	13.	1.5	1.	5.0	0.10
New brand (d)	17.	2.0	1.	5.5	0.10

D. Time in years required for market introduction:

Brand	100-Percent Freeze-Dried	20–80 Mix
Maxwell House	4.0	3.0
Yuban	2.5	2.0
Super Yuban	2.5	2.0
New brand (c)	2.5	2.0
New brand (d)	2.5	2.0

E. Decimals of first brand's sales volume expected to come from Maxwell House Division ground and soluble coffees:

EXHIBIT 10–1 (Continued)

First Brand	Ground	Soluble
Maxwell House	.20 [c]	.10 [c]
Yuban	.10 [c]	.15 [c]
Super Yuban	.15	.30
New brand (c)	.10	.25
New brand (d)	.15	.35

[c] Over base volume

F. Decimals of second brand's sales volume expected to come from Maxwell House Division ground, spray-dried, and freeze-dried coffees, given the identity of the first brand:

First Brand	Second Brand	Ground	Spray-Dried	Freeze-Dried
Maxwell House	Yuban	.05	.50	.05
	Super Yuban	.12	.20	.10
	New brand (c)	.12	.05	.25
	New brand (d)	.15	.15	.20
Yuban	Maxwell House	.02	.82	.02
	New brand (c)	.10	.25	.10
	New brand (d)	.15	.35	.05
Super Yuban	Maxwell House	.05	.80	.02
	New brand (c)	.10	.25	.10
	New brand (d)	.15	.35	.05
New brand (c)	Maxwell House	.03	.75	.05
	Yuban	.05	.60	.05
	Super Yuban	.10	.40	.05
	New brand (d)	.15	.35	.10
New brand (d)	Maxwell House	.03	.85	0
	Yuban	.05	.65	0
	Super Yuban	.10	.40	0
	New brand (c)	.13	.25	0

G. Years in which a freeze-dried coffee marketed by another company might first affect Maxwell House Division coffee sales and the probabilities for those years:

Year	Probability
1968	.25
1969	.45
1970	.30

H. Decimal of original (first-brand) sales-volume estimate applicable if brand is second in order, given identity of first freeze-dried brand and assuming no outside freeze-dried-coffee competition:

EXHIBIT 10–1 (Continued)

First Brand	Second Brand	Decimal
Maxwell House	Yuban	.82
	Super Yuban	.85
	New brand (c)	.60
	New brand (d)	1.00
Yuban	Maxwell House	.95
	New brand (c)	.90
	New brand (d)	1.00
Super Yuban	Maxwell House	.97
	New brand (c)	.93
	New brand (d)	1.00
New brand (c)	Maxwell House	.88
	Yuban	.82
	Super Yuban	.85
	New brand (d)	1.00
New brand (d)	Maxwell House	.94
	Yuban	.84
	Super Yuban	.87
	New brand (c)	.85

I. Decimal of original (first-brand) sales-volume estimate applicable as average for period after outside freeze-dried-coffee competition has first affected Maxwell House Division coffee sales:

Brand	Decimal
Maxwell House	.85
Yuban	.80
Super Yuban	.70
New brand (c)	.50
New brand (d)	1.00

J.

1.	Depreciation rate for declining-balance method	.20
2.	Estimated average gross margin for Maxwell House Division ground coffees minus average marketing cost (expressed in dollars per unit)	1.05
3.	Estimated average gross margin for Maxwell House Division soluble coffees minus average marketing cost (expressed in dollars per unit)	1.50
4.	SALFCT (a factor that will be multiplied by net plant to obtain market value of plant at end of last year in period of analysis)	1.00
5.	Discount rate (expressed as a decimal)	.09
6.	Number of years chosen for time period of analysis	15.00
7.	Probability of successfully developing a 20–80 mix	.25
8.	Rate of return chosen to be high enough to exceed the actual rate of return for all alternatives	.30
9.	Code to indicate whether one-brand or two-brand alternatives are to be analyzed (1 = one brand; 0 = two-brand sequence)	0

EXHIBIT 10–1 (Continued)

K. Probabilities of sales estimates for second brand, conditional on sales volume of first brand:

Sales Volumes for First Brand	Sales Estimates for Second Brand: Highest	2d Highest	3d Highest	Lowest
Maxwell House		Yuban		
Highest	.20	.35	.35	.10
2d highest	.10	.30	.40	.20
3d highest	.05	.20	.50	.25
Lowest	.00	.15	.45	.40
Maxwell House		Super Yuban		
Highest	.15	.20	.35	.30
2d highest	.10	.10	.30	.50
3d highest	.05	.10	.30	.55
Lowest	.00	.05	.35	.60
Maxwell House		New brand (c)		
Highest	.20	.25	.35	.20
2d highest	.10	.20	.40	.30
3d highest	.05	.15	.45	.35
Lowest	.00	.10	.50	.40
Maxwell House		New brand (d)		
Highest	.15	.45	.35	.05
2d highest	.10	.40	.40	.10
3d highest	.05	.35	.35	.25
Lowest	.05	.25	.30	.40
Yuban		Maxwell House		
Highest	.25	.55	.15	.05
2d highest	.20	.55	.15	.10
3d highest	.20	.50	.20	.10
Lowest	.10	.40	.30	.20
Yuban		New brand (c)		
Highest	.20	.30	.30	.20
2d highest	.10	.25	.40	.25
3d highest	.10	.20	.40	.30
Lowest	.05	.15	40	.40
Yuban		New brand (d)		
Highest	.15	.45	.30	.10
2d highest	.10	.40	.40	.10
3d highest	.10	.40	.40	.10
Lowest	.05	.35	.35	.25
Super Yuban		Maxwell House		
Highest	.25	.60	.15	.0
2d highest	.20	.60	.15	.05
3d highest	.20	.55	.15	.10
Lowest	.15	.45	.25	.15

EXHIBIT 10–1 (Continued)

Sales Volumes for First Brand	Sales Estimates for Second Brand			
	Highest	2d Highest	3d Highest	Lowest
Super Yuban		New brand (c)		
Highest	.20	.30	.30	.20
2d highest	.15	.25	.35	.25
3d highest	.10	.20	.45	.25
Lowest	.05	.15	.45	.35
Super Yuban		New brand (d)		
Highest	.20	.45	.30	.05
2d highest	.15	.45	.30	.10
3d highest	.10	.45	.35	.10
Lowest	.10	.35	.40	.15
New brand (c)		Maxwell House		
Highest	.35	.55	.10	.0
2d highest	.25	.60	.15	.0
3d highest	.20	.50	.20	.10
Lowest	.10	.30	.35	.25
New brand (c)		Yuban		
Highest	.30	.45	.25	.0
2d highest	.20	.35	.35	.10
3d highest	.10	.30	.40	.20
Lowest	.0	.20	.40	.40
New brand (c)		Super Yuban		
Highest	.20	.30	.30	.20
2d highest	.15	.20	.40	.25
3d highest	.10	.10	.30	.50
Lowest	.00	.05	.25	.70
New brand (c)		New brand (d)		
Highest	.25	.50	.25	.0
2d highest	.20	.50	.25	.05
3d highest	.10	.40	.40	.10
Lowest	.05	.20	.35	.40
New brand (d)		Maxwell House		
Highest	.25	.55	.15	.05
2d highest	.20	.50	.20	.10
3d highest	.20	.50	.20	.10
Lowest	.15	.40	.30	.15
New brand (d)		Yuban		
Highest	.15	.35	.35	.15
2d highest	.10	.30	.40	.20
3d highest	.10	.30	.40	.20
Lowest	.05	.20	.50	.25
New brand (d)		Super Yuban		
Highest	.15	.15	.35	.35
2d highest	.10	.10	.30	.50
3d highest	.10	.10	.30	.50
Lowest	.05	.05	.25	.65

EXHIBIT 10–1 (Concluded)

Sales Volumes for First Brand	Sales Estimates for Second Brand			
	Highest	2d Highest	3d Highest	Lowest
New brand (d)		New brand (c)		
Highest	.15	.30	.35	.20
2d highest	.10	.20	.40	.30
3d highest	.10	.20	.40	.30
Lowest	.05	.15	.40	.40

EXHIBIT 10–2

TWO-BRAND ALTERNATIVES: ACCUMULATED PRESENT VALUE OF EXPECTED CASH FLOW BASED ON TEN-YEAR AND FIFTEEN-YEAR PERIODS OF ANALYSIS
(Millions of Dollars)

Two-Brand Sequence	Try to Develop 100-Percent Freeze-Dried Coffee First		Try to Develop 20–80 Mix First	
	10 Years	15 Years	10 Years	15 Years
Maxwell House–Yuban	−42.59	−46.83	−20.90	−20.74
Maxwell House–Super Yuban	−35.58	−35.64	−16.25	−12.13
Maxwell House–New brand (c) [a]	−36.32	−35.46	−17.02	−11.94
Maxwell House–New brand (d) [a]	−34.26	−22.11	−12.97	5.19
Yuban–Maxwell House	−41.61	−48.04	−20.19	−22.44
Yuban–New brand (c)	−13.44	− 4.51	− 3.20	6.60
Yuban–New brand (d)	−10.55	7.88	0.02	22.00
Super Yuban–Maxwell House	−33.02	−34.85	−13.81	−11.73
Super Yuban–New brand (c)	− 5.55	7.42	2.79	16.26
Super Yuban–New brand (d)	− 2.09	21.07	6.15	32.19
New brand (c)–Maxwell House	−33.60	−29.76	−11.32	− 3.16
New brand (c)–Yuban	−12.26	− 0.93	− 0.01	12.22
New brand (c)–Super Yuban	− 4.43	11.15	5.58	21.94
New brand (c)–New brand (d)	− 3.86	25.84	9.38	42.84
New brand (d)–Maxwell House	−31.42	−17.72	− 7.70	12.71
New brand (d)–Yuban	− 8.77	11.16	2.95	27.00
New brand (d)–Super Yuban	− 1.48	22.91	8.16	36.31
New brand (d)–New brand (c)	− 3.90	24.42	8.93	41.04

[a] The letters (c) and (d) after "New brand" stand for "caffein" and "decaffeinated," respectively.

EXHIBIT 10–3

TWO-BRAND ALTERNATIVES: RATE OF RETURN ON INVESTMENT AND PROFITABILITY-INDEX VALUE BASED ON FIFTEEN-YEAR PERIOD OF ANALYSIS

Two-Brand Sequence	Try to Develop 100-Percent Freeze-Dried Coffee First		Try to Develop 20–80 Mix First	
	ROFRT[c]	EPINX[d]	ROFRT[c]	EPINX[d]
Maxwell House–Yuban	NPR[a]	0.525	0.004	0.710
Maxwell House–Super Yuban	NPR	0.622	0.038	0.823
Maxwell House–New brand (c)[b]	NPR	0.636	0.041	0.830
Maxwell House–New brand (d)[b]	0.045	0.831	0.106	1.056
Yuban–Maxwell House	NPR	0.494	NPR	0.673
Yuban–New brand (c)	0.070	0.915	0.139	1.190
Yuban–New brand (d)	0.114	1.096	0.194	1.387
Super Yuban–Maxwell House	NPR	0.615	0.034	0.821
Super Yuban–New brand (c)	0.124	1.155	0.227	1.528
Super Yuban–New brand (d)	0.158	1.281	0.258	1.617
New brand (c)–Maxwell House	0.016	0.736	0.078	0.960
New brand (c)–Yuban	0.087	0.985	0.167	1.302
New brand (c)–Super Yuban	0.131	1.192	0.231	1.583
New brand (c)–New brand (d)	0.148	1.246	0.245	1.613
New brand (d)–Maxwell House	0.055	0.877	0.127	1.124
New brand (d)–Yuban	0.121	1.125	0.209	1.438
New brand (d)–Super Yuban	0.156	1.268	0.256	1.618
New brand (d)–New brand (c)	0.148	1.235	0.251	1.588

[a] NPR stands for "no positive rate of return."
[b] The letters (c) and (d) after "New brand" stand for "caffein" and "decaffeinated," respectively.
[c] Internal rate of return on investment in percent per year.
[d] Profitability index.

EXHIBIT 10–4

THE TOP EIGHT TWO-BRAND ALTERNATIVES RANKED BY MEASURES FROM PRIOR ANALYSIS BASED ON FIFTEEN-YEAR TIME PERIOD AND DECISION TO TRY FIRST TO DEVELOP 20–80 MIX

	Accumulated Present Value of Expected Cash Flow		Rate of Return		Profitability Index	
Two-Brand Sequence	Rank	Millions of Dollars	Rank	Per-cent	Rank	Value
New brand (c)–New brand (d) [a]	1	42.84	4	24.5	3	1.613
New brand (d)–New brand (c)	2	41.04	3	25.1	4	1.588
New brand (d)–Super Yuban	3	36.31	2	25.6	1	1.618
Super Yuban–New brand (d)	4	32.19	1	25.8	2	1.617
New brand (d)–Yuban	5	27.00	7	20.9	7	1.438
Yuban–New brand (d)	6	22.00	8	19.4	8	1.387
New brand (c)–Super Yuban	7	21.94	5	23.1	5	1.583
Super Yuban–New brand (c)	8	16.26	6	22.7	6	1.528

[a] The letters (c) and (d) after "New brand" stand for "caffein" and "decaffeinated," respectively.

EXHIBIT 10–5

EXPOSURE TO GAIN AND LOSS OF THE TWO-BRAND ALTERNATIVES RANKING HIGHEST ON ACCUMULATED PRESENT VALUE OF EXPECTED CASH FLOW AND RATE OF RETURN ON INVESTMENT

	Try to Develop 100-Percent Freeze-Dried Coffee First		Try to Develop 20–80 Mix First	
Two-Brand Sequence	APVECF [a]	Probability	APVECF	Probability
Based on 15-Year Period of Analysis				
New brand (c) [b]– New brand (d) [b]	44	.100	171	.025
	35	.200	145	.050
	25	.400	117	.100
	15	.300	91	.075
			30	.075
			24	.150
			17	.300
			9	.225
New brand (d)–New brand (c)	35	.100	137	.025
	29	.400	122	.100
	21	.400	108	.100
	12	.100	89	.025
			24	.075
			20	.300
			14	.300
			7	.075
New brand (d)–Super Yuban	33	0.100	118	0.025
	27	0.400	105	0.100
	19	0.400	90	0.100
	11	0.100	73	0.025
			23	0.075
			19	0.300
			13	0.300
			7	0.075
Super Yuban–New brand (d)	32	0.100	116	0.025
	26	0.100	102	0.025
	21	0.300	89	0.075
	18	0.500	77	0.125
			21	0.075
			17	0.075
			14	0.225
			11	0.375

[a] APVECF stands for "accumulated present value of expected cash flow in millions of dollars."

[b] The letters (c) and (d) after "New brand" stand for "caffein" and "decaffeinated," respectively.

EXHIBIT 10–5 (Continued)

Two-Brand Sequence	Try to Develop 100-Percent Freeze-Dried Coffee First		Try to Develop 20–80 Mix First	
	APVECF [a]	Probability	APVECF	Probability
Based on 10-Year Period of Analysis				
New brand (c)–New brand (d)	2	.100	86	.025
	−1	.200	71	.050
	−4	.400	55	.100
	−8	.300	40	.075
			−4	.075
			−5	.150
			−6	.300
			−8	.225
New brand (d)–New brand (c)	−0	.100	63	.025
	−2	.400	55	.100
	−5	.400	48	.100
	−9	.100	39	.025
			−3	.075
			−4	.300
			−6	.300
			−8	.075
New brand (d)–Super Yuban	2	.100	53	.025
	0	.400	47	.100
	−3	.400	40	.100
	−6	.100	31	.025
			−1	.075
			−2	.300
			−4	.300
			−6	.075
Super Yuban–New brand (d)	1	.100	53	.025
	0	.100	45	.025
	−2	.300	38	.075
	−3	.500	31	.125
			−2	.075
			−3	.075
			−4	.225
			−5	.375

EXHIBIT 10-6

PRIOR ANALYSIS PRINT-OUT FOR NEW BRAND (C)—NEW BRAND (D) SEQUENCE ON 20–80 MIX

FIRST BRAND INTRODUCED 4 SECOND BRAND INTRODUCED 5 MIX 2

YEAR	1ST BRAND SALES (MILLIONS OF UNITS)	2ND BRAND SALES (MILLIONS OF UNITS)	1ST BRAND CANIBALISM (MILLIONS OF $)	2ND BRAND CANIBALISM (MILLIONS OF $)	PLANT INVESTMENT (MILLIONS OF $)	NET PLANT	CASH FLOW	PV CASH FLOW	COMBINED CASH FLOW	COMBINED PV CASH FLOW
1962	0.00	0.00	0.0	0.0	0.0	0.0	0.0	0.0	0.0	0.0
1963	0.00	0.00	0.0	0.0	1.1	1.0	-1.1	-0.9	-0.3	-0.2
1964	0.67	0.00	0.3	0.0	8.8	8.8	-9.2	-7.1	-2.3	-1.8
1965	5.29	0.62	2.5	0.4	15.0	20.5	-12.4	-8.8	-5.5	-3.9
1966	9.53	4.97	4.6	3.4	9.7	25.2	3.1	2.0	-18.8	-12.2
1967	10.58	9.32	5.1	6.4	2.3	22.2	20.6	12.3	-21.7	-13.0
1968	10.04	10.70	3.6	5.5	1.8	19.4	25.5	13.9	-2.0	-1.1
1969	7.92	11.79	1.1	2.4	2.0	17.4	26.6	13.4	15.6	7.8
1970	6.42	13.00	0.0	0.0	2.2	15.9	27.9	12.8	19.3	8.9
1971	7.07	14.34	0.0	0.0	2.4	14.9	30.4	12.9	20.3	8.6
1972	7.80	15.81	0.0	0.0	2.7	14.4	33.3	12.9	21.5	8.3
1973	8.60	17.43	0.0	0.0	3.0	14.2	36.5	13.0	23.1	8.2
1974	9.48	19.21	0.0	0.0	4.1	15.0	39.3	12.8	22.0	7.2
1975	10.45	21.18	0.0	0.0	5.8	17.2	42.1	12.6	23.1	6.9
1976	11.52	23.35	0.0	0.0	6.4	19.5	46.4	12.8	25.4	7.0

INDEXES FOR EXPECTED OUTCOME
ACCUMULATED PRESENT VALUE INDEX 42.84
PROFITABILITY INDEX 1.613
INTERNAL RATE OF RETURN 0.245

OUTCOME	EXPOSURE TO RISK	PROBIBILITY OF OCCURANCE
1	171.	0.025
2	145.	0.050
3	117.	0.100
4	91.	0.075
5	30.	0.075
6	24.	0.150
7	17.	0.300
8	9.	0.225

Note: "Cash Flow" and "PV Cash Flow" assume success in developing 20–80 mix. "Combined Cash Flow" and "Combined PV Cash Flow" represent expected values which allow for probabilities of success and failure on an attempt to develop 20–80 mix.

EXHIBIT 10–7

ESTIMATED PLANT CAPACITY IN EXCESS OF SALES VOLUME, BY YEARS FOR THE FOUR LEADING TWO-BRAND ALTERNATIVES[a]
(Millions of Units)

Year	New Brand (c)– New Brand (d)	New Brand (d)– New Brand (c)	New Brand (d)– Super Yuban	Super Yuban– New Brand (d)
1962	0.00	0.00	0.00	0.00
1963	0.00	0.00	0.00	0.00
1964	0.00	0.00	0.00	0.00
1965	0.00	0.00	0.00	0.00
1966	0.00	0.00	0.00	0.00
1967	0.00	0.00	0.00	0.00
1968	0.54	0.00	0.00	0.22
1969	2.66	1.51	0.35	0.71
1970	4.16	2.55	0.56	1.02
1971	3.51	1.96	0.36	0.84
1972	2.78	1.30	0.15	0.66
1973	1.98	0.58	0.00	0.47
1974	1.10	0.00	0.00	0.27
1975	0.13	0.00	0.00	0.06
1976	0.00	0.00	0.00	0.00
Total	16.86	7.90	1.42	4.25

[a] Figures are based on use of 20–80 mix and fifteen-year period of analysis.
[b] The letters (c) and (d) after "New brand" stand for "caffein" and "decaffeinated," respectively.

EXHIBIT 10–8

ONE-BRAND ALTERNATIVES: ACCUMULATED PRESENT VALUE OF EXPECTED CASH FLOW, RATE OF RETURN ON INVESTMENT, AND PROFITABILITY-INDEX VALUE, BASED ON FIFTEEN-YEAR PERIOD OF ANALYSIS

	100-Percent Freeze-Dried Coffee			20–80 Mix		
Brand	APVECF[a]	ROFRT[b]	EPINX[c]	APVECF	ROFRT	EPINX
Maxwell House	−33.48	NPR[d]	0.536	−22.31	NPR	0.660
Yuban	− 9.59	NPR	0.601	−5.68	0.015	0.724
Super Yuban	3.44	0.131	1.191	5.20	0.170	1.344
New brand (c)[e]	8.96	0.133	1.198	12.12	0.166	1.322
New brand (d)[e]	21.15	0.164	1.294	25.32	0.200	1.406

[a] APVECF stands for "accumulated present value of expected cash flow in millions of dollars."
[b] ROFRT stands for internal rate of return in percent per year.
[c] EPNIX stands for profitability index.
[d] NPR stands for "no positive rate of return."
[e] The letters (c) and (d) after "New brand stand for "caffein" and "decaffeinated," respectively.

Section Three

Concluding Comments on Applying the Bayesian Approach

CHAPTER 11

SUMMARY OBSERVATIONS ON MANAGEMENT APPLICATIONS OF DECISION THEORY

The work of this book was undertaken to learn more about what is involved in employing the Bayesian approach in realistic business management contexts. The main aspects of such applications were covered in the preceding chapters. Summary observations and concluding comments on the requirements, limitations, and potential benefits of using the approach are presented here.

The decision problems chosen for special attention illustrated various elements of decision making. They included identifying and screening alternative courses of action, structuring problems in the form of decision trees, specifying measures to be used in evaluating alternatives, planning the analysis, identifying and developing the inputs needed for the computations, deciding whether more information should be sought before choosing among the alternative actions, evaluating research proposals, incorporating the findings of market tests and consumer surveys into the analysis so that they might have appropriate influence, summarizing and presenting the output of the analysis, and making a final choice.

MANAGEMENT DECISIONS

The quick-strip-can and the freeze-dried-coffee problems served to emphasize the element of uncertainty that pervades important management decision making. Although past experience may be helpful in thinking about the problem at hand, it typically falls far short of providing a record that directly indicates the possible outcomes and their probabilities for a course of action under consideration. In addition, the alternatives that might be followed frequently are not immediately apparent. They remain to be identified and formulated.

To a marked extent, important management decisions are unique. This was seen in both the quick-strip-can and the freeze-dried-coffee illustrations. The former had more background information because a major change of container had been made recently. The outcome of that action and the results of related research were available. Even so, they were of limited help because of the newness of the quick-strip feature and its higher cost, which appeared to require a price increase. Freeze-dried coffee was an innovation, so there was little in the way of empirical evidence for use in estimating the probable demand for different brand concepts at different prices. And demand, of course, depended in part on the unknown future actions of competitors.

Notwithstanding such uncertainty, actions must be chosen. The question is how this can best be done under the circumstances. Different approaches have been advocated over the years.[1] A number of statisticians have been unwilling to admit managerial judgment to formal analyses, holding that meaningful probabilities can not be assigned to possible events and consequences of actions under uncertainty. Decision criteria advocated by these persons have tended to be conservative, for example, choosing the action with the highest minimum payoff (sometimes with and sometimes without taking into account lost opportunity for profits from failure to take the right action at the right time). Some have advocated using the criterion of expected value but with the stipulation that it be computed after assigning equal probabilites to all possible events in a set. Such approaches are unacceptable in the business world, where managerial experience and judgment are regarded as primary assets that should play a dominant role in decision making.

The prevailing business practice has been to draw on experience and judgment informally. When formal analyses have been undertaken, one best estimate frequently has been used as a certainty equivalent, thus playing down the uncertainty present. A variation consists of adding maximum and minimum estimates without assigning probabilities to them, a step which recognizes that the outcomes can vary but does not formally attempt to measure the uncertainty.

In contrast, the Bayesian approach provides for considering the effects of alternative chance events on the outcomes of interest, makes uncertainty explicit in the form of subjective probability distributions, and helps the decision maker to be consistent in his reasoning. Through the

[1] For a historical review, see Howard Raiffa, *Decision Analysis: Introductory Lectures on Choices Under Uncertainty* (Reading, Mass.: Addison-Wesley Publishing Co., Inc., 1969), chapter 10, pp. 273–300. For a description of the maximin, minimax regret, and the Laplace criteria, see Wroe Alderson and Paul E. Green, *Planning and Problem Solving in Marketing* (Homewood, Ill.: Richard D. Irwin, Inc., 1964), pp. 88–93.

measure of expected value, it provides for evaluating the gambles that are represented by the alternative courses of action. The purpose is to enable the decision maker to recognize and choose "good bets" in terms of his judgments and preferences.[2] The extent to which the objective will be realized in practice depends on a number of factors already considered in this book and which will receive additional discussion.

COMMENTS ON APPLYING BAYESIAN PRIOR ANALYSIS

The Bayesian approach represents a way of thinking that is structured, probabilistic, and specific. Applications can vary, however, in the extent to which they represent these qualities. The possibilities range from the sketching of a limited decision tree without actually undertaking calculations to the carrying out of a complex, detailed analysis by computer. The requirements of the approach will vary accordingly, something that should be kept in mind while reading the comments that follow, which assume, for the most part, reasonably full implementation.

RECOGNITION OF UNCERTAINTY. By its nature, the Bayesian approach requires that uncertainty be recognized. This point deserves special mention because such direct recognition is contrary to much business practice. It has been common, for example, to expect "good executives" to know "for sure"; and many executives have risen to power by persuasively assuming a posture of certainty about the uncertain future outcomes of the actions they advocated. Convention, therefore, may constitute a source of resistance to the Bayesian approach.

CLARIFICATION OF DECISION CRITERIA. Effective use of decision analysis depends on clear identification of the criteria that are to be used in choosing among the alternatives. In outlining the approach, this book advocated the use of (1) expected value of one or more measures of payoff and (2) measures of exposure to gain and loss. The first is an expression of value of the gamble represented by a course of action; whereas the second portrays the risk associated with it.

Much of the literature on Bayesian decision theory has assumed the objective of maximizing monetary payoff and has used a single measure, usually revenue or profit, in simple illustrations. Actually, of course, there are a number of measures of monetary consequences (profit, cash flow, rate of return, profitability index, and so forth), and major business decisions typically require that attention be paid to more than one of them. A determination must be made concerning which measure or measures are most suitable and, if more than one is to be used, how they are to be viewed in combination in the reaching of a decision. These

[2] For a clear treatment of the rationale involved, see Howard Raiffa, *op. cit.*, chapter 5, "The Use of Subjective Probability," pp. 104–128.

considerations received attention in the analysis of the freeze-dried-coffee alternatives. The procedure used was to compute expected values of each of several measures and present them to the decision maker, who could view them as separate entities and, if he wished, combine them by using a weighting scheme reflecting his judgment of their relative importance.

The decision maker may also wish to consider expected values of measures that are not strictly monetary. In the quick-strip-can problem, for example, share of market was important because of both its monetary and its nonmonetary implications. Its relevance to profit in the long run is indicated by the large amounts of effort and dollars spent by business to increase market share. In addition to the strictly profit and loss consequences, share of market frequently serves as an indicator of a company's position of leadership and a manager's effectiveness. These are considerations that an individual decision maker must keep in mind, especially in organizations that have great pride in their leadership and that emphasize growth. In such an environment, a loss in share, other than a very small one, might not be tolerated. Even if it were, it could draw unwanted skeptical attention to the manager in charge.

One way of handling such considerations with the interest of the manager paramount was illustrated in the quick-strip-can analysis. It provided for adjusting monetary values to reflect the executive's assessment of implications of gains and losses in share of market not fully taken into account in a short-run financial analysis. His responses to certain questions served as a basis for assigning to the possible changes in share of market dollar values that presumably took into account all considerations he regarded as relevant. The values represented a scale of the relative attractiveness to the decision maker of the various outcomes.

The procedure just described is one of several that can be used for arriving at expected values which reflect both the monetary value of a given course of action and the decision maker's attitude toward financial risk (see references to utility theory in Chapter 1, page 9). Another approach consists of asking a series of questions, usually in the form of posing hypothetical bets, for the purpose of constructing a utility curve (a scale) that measures the decision maker's preference for gambles with different payoffs subject to risk. The curve provides a means of translating dollar consequences into the scale values. Expected utility (the expected value of the scale value) is computed for each of the alternatives being evaluated. The decision rule is to choose the action with the highest expected utility.

Although utility measures that theoretically reflect attitude toward risk can be constructed, whether they add anything of practical value to the making of business decisions of major importance is an open question. Utility (and perhaps scale values in general) is not familiar as an operating concept to many business executives. They are accustomed to thinking in terms of one or more measures of monetary outcome and the odds of realizing the latter in specific situations. If these terms are not employed in the questioning procedures used to derive utility measures, such measures may not accurately or fully express a decision maker's attitude toward the risk that is in the particular actions being evaluated.

Certain psychologists have questioned the adequacy of the concept of maximizing expected utility for allowing for perceived risk and are working to develop an alternative theory relative to risky decisions.[3] Procedures for dealing with nonmonetary outcome measures are in an early stage of development and are controversial. For practical and theoretical reasons, therefore, one can argue that currently there is no better way to serve the decision maker than to present him with both (1) expected values of the appropriate monetary consequences and (2) measures of the exposure to gain and loss for each alternative, and then to allow him to examine them, applying his own definition of risk in his own way.

FORMULATING AND SCREENING OF ALTERNATIVES. Decision theory is concerned with evaluating alternatives once they have been formulated. Considering the nature of the Bayesian approach, however, one might expect its use to indirectly benefit the work that must precede formal analysis. For example, the demand for specific alternatives might stimulate the creative process, so that a greater than usual number of attractive courses of action come into being. In addition, they might tend to be more fully developed. The screening of many alternatives to a large extent must be an informal process, and there is no general rule about how many possible actions should be submitted to formal evaluation. The screening, however, might be facilitated by a greater awareness of important criteria than would have been present without the influence of the rigors of Bayesian analysis. Whether such potential benefits will be realized in practice would appear to depend on whether the same people are involved in the formulation and screening as in the evaluation and, if not, the amount and character of communication that takes place between the different parties.

HANDLING OF COMPLEXITY. The Bayesian approach and the computer make possible detailed analyses of complex decision problems. The former provides for considering the effects of alternative chance events, whereas the computer can quickly dispose of the large number of computations involved. No one has to keep everything in mind at one time. The structuring of the problem separates the relevant factors so that parts of the problem can be worked on by different persons accord-

[3] For example, see Clyde H. Coombs and Lily C. Huang, "A Portfolio Theory of Risk Preference," Working Paper 68–5 of the Michigan Mathematical Psychology Program, Department of Psychology, The University of Michigan, Ann Arbor, Mich., 1968.

ing to their qualifications. Sequential aspects of decision making can be handled in an orderly manner. In the quick-strip-can decision problem, for example, a course of action was analyzed that consisted of placing the new container on the market but later abandoning it if sales did not meet a specified level in a given period of time.

Although complexity may not be confined by the capacity of the computer, there are other limitations. Certainly one should not try to break down the problem so finely that it becomes unmanageable. The amount of detail need not exceed that required by the decision-making objective. Unfortunately, nothing very specific can be said at this time about how to recognize the optimum amount of detail—to a marked extent this is an art.

MAKING ASSUMPTIONS EXPLICIT. The main assumptions to be used are recorded or reflected in the model itself or in the input data. This feature has several implications. On the negative side, some executives may resist participation because they prefer to avoid what they see as a personal risk in such an exposure of their assumptions. For the users, however, there are potential benefits. The approach brings rigor that is conducive to thorough consideration of the elements of the problem. Its explicitness gives visibility to the informational requirements of the analysis that should lead to greater than usual consideration of undertaking research to aid the decision making. It may also lead to improved research design because of a sharper definition of the informational objectives. Another potential advantage is improved communications among executives. Differences of opinion can be pinpointed quickly, and discussions can proceed with a clearer understanding of the positions of the participants and a saving of time.

OBTAINING GOOD SUBJECTIVE PROBABILITIES. The value of the Bayesian approach is dependent on using subjective probabilities that reflect with reasonable accuracy the basic judgments of the decision maker. The need for a sound procedure for eliciting the probabilities was noted earlier. There is no standard approach. Instead, several have been employed. One of them, which divides the range of possibilities into quadrants of probability, was described in Chapter 4. Another consists of having the decision maker draw a curve for the cumulative distribution and then subdivide it into segments for which discrete values are obtained for use in calculating expected value.[4] Relatively little is known at the present time about the effect of differences in procedure on the values and the stability of the resulting subjective probabilities. Earlier, it was pointed out that in theory the use of a probability distribution

[4] For a more detailed discussion of assessment of subjective probabilities, see Robert Schlaifer, *Analysis of Decisions Under Uncertainty* (New York: McGraw-Hill Book Company, 1969), chapter 8, pp. 280–317.

normally is superior to the use of one best estimate. Little empirical evidence has been gathered to date, however, about what difference the choice makes under different conditions in business practice.

If a decision maker consults more than one source about the effect of one or more factors relevant to the analysis, he must decide what to do about the several sets of judgments. If he elects to be influenced by more than one set, he faces the task of converting them into a single set of probability distributions. He may do this intuitively, or he can employ a formal procedure for giving what he regards as appropriate weights to the opinions of the different sources. Whether a formal approach is better than intuition is an open question that should be considered in a specific context. Methods for evaluating and combining subjective judgments have received some attention in the literature.[5]

INDIVIDUAL VERSUS GROUP DECISION MAKING. Most of the literature on Bayesian decision analysis, including this book, has focused on the process by which an individual arrives at his own position concerning what choice should be made. In so doing, he is able to use whatever information and opinions of others he chooses, but no group consensus is required. The literature is also applicable to decisions made by more than one individual, providing that the participants have common objectives, decision criteria, and attitudes toward risk and that they can agree on the assumptions and input data to be used in the analysis—in other words, if they behave like a single individual.

Groups that have decision-making responsibility seldom are that homogeneous. The process of group decision making, therefore, is more complex because it must provide not only for the functions discussed in the context of individual decision making but also for the resolution of differences among its members. The latter is in the realm of group dynamics and is beyond the scope of Bayesian decision theory. The business literature does not yet include reports of attempts to use the Bayesian approach in group decision making. Most of the comments made here about the approach appear to be applicable in the group context. Use of the approach can be expected to make more apparent than usual any differences that are present in the group on decision criteria, risk aversion, judgments, and objectives—factors that often are related to personal ambitions of executives who may be in competition with one another. Whether greater exposure of latent conflict and its sources would result in explosion or facilitate reaching agreement would appear to depend on organizational, personal, temporal, and other considerations, some of which may not be known at the present time.

[5] See Robert L. Winkler, "The Consensus of Subjective Probability Distributions," *Management Science* (October, 1968) : B61–B75; and G. J. Brabb and E. D. Morrison, "The Evaluation of Subjective Information," *Journal of Marketing Research* Vol. 1, no. 4 (November 1964) : 40–44.

SENSITIVITY ANALYSIS. The Bayesian approach, when employed with the computer, facilitates the testing of the sensitivity of outcomes to changes in assumptions represented in the input data. Once factors have been identified as critical, attention can be focused on considering what, if anything, can be done to improve their estimates. Concern over the estimates for other factors is eased. It is not unusual to find that the values for some variables can change greatly without significantly affecting the outcome. The use of sensitivity analysis was illustrated in the freeze-dried-coffee analysis in regard to time period. Choosing the time period often is a difficult decision, and there is no objective procedure for doing it. The longer the period, the greater the uncertainty of outcomes, but the effects of actions being considered may extend far into the future. One would like to go far enough to encompass effects of actions that are large enough to make a difference in which alternative would rank first. It was helpful in the freeze-dried-coffee problem to learn that the top rankings were not very sensitive to a change from ten to fifteen years.

COST. The cost of employing Bayesian analysis can vary considerably depending on the complexity of the problem and whether applicable computer programs are on hand or have to be developed. The time required for building and programming the analytical model is the critical element of expense. The cost of computations by computer is low and declining. The modeling and programming for an analysis of the complexity seen in the freeze-dried-coffee decision can take several months. This may or may not be more time than a given organization would spend on such a problem under a different decision-making approach, whether that approach is formal or informal. If a decision must be made quickly, however, a detailed formal analysis is not possible. The time requirement of the modeling and programming for the quick-strip-can prior and preposterior analyses was about one-fifth as great as that of the freeze-dried-coffee prior analysis. The preposterior analysis program is one that can be used or easily adapted for use for other preposterior analyses in the future.

COMMENTS ON APPLYING BAYESIAN PREPOSTERIOR ANALYSIS

A most important contribution of Bayesian statistics is that it provides a formal analytical framework for evaluating research. Its use has the potential for great improvement in business decisions on research proposals. Such decisions too often have been swayed by a clear view of direct costs while opportunity costs and the value of new information that might result were never brought into focus. Research frequently has been ordered when it should not have been and omitted when the situation warranted a substantial research expenditure.

The use of preposterior analysis forces managers and researchers to think of both the costs and the value of proposed research in terms of reducing the costs of wrong decisions. Such consideration should help guard against undertaking research when prior uncertainty and the cost of making a wrong decision both are low, when the research is unlikely to yield information that would reduce uncertainty enough to lead to a different choice than would have been made without it, and when opportunity cost of delay is prohibitive. By making opportunity cost apparent, preposterior analysis guards against neglecting this element, which frequently is of far greater consequence than the direct costs of research. The use of preposterior analysis should encourage large research expenditures when they are justified by the importance of the decision, the uncertainty in the decision situation, and the prospects of the research producing information that will reduce uncertainty.

Other implications of the use of Bayesian preposterior analysis will be noted in the discussion of its requirements.

EVALUATION OF INFORMATION. Bayesian preposterior analysis requires that the decision maker (or someone to whom the responsibility is delegated) evaluate the information that research might produce in terms of change in the prior probabilities. In so doing, he must allow for the probability that a given research observation might be made under several different states of nature.

A consequence of the necessity of evaluating research observations is greater recognition of the sources of error that may be present and consideration of making allowances for them.[6] This subject was given attention in the discussion of the quick-strip-can preposterior analysis. In employing market and consumer-use tests and surveys, business typically has given substantial attention to sampling errors while neglecting measurement and prediction errors. Test marketing, for example, has been widely used even though evidence on the projectability of the results has been limited. The need for empirical research was highlighted in a study by Gold,[7] which found test-marketing prediction errors to be large. The limitations of such tests and the need for controlled experimental research have received increased attention in recent years.[8]

[6] See Rex V. Brown, *Research and the Credibility of Estimates* (Boston: Division of Research, Harvard Business School, 1969); idem, "Evaluation of Total Survey Error," *Journal of Marketing Research* Vol. 4, no. 2 (May 1967):117–127.

[7] Jack A. Gold, "Testing Test Market Predictions," *Journal of Marketing Research* Vol. 1, no. 3 (August 1964):8–16.

[8] For example, see Frank Stanton, "What Is Wrong with Test Marketing?" *Journal of Marketing* Vol. 31, no. 2 (April 1967):43–47; and David K. Hardin, "A New Approach to Test Marketing," *Journal of Marketing* Vol. 30, no. 4 (October 1966):28–31.

Use of the Bayesian approach should encourage better recording of experience by business as well as research designed to learn more about the meanings of different research observations made under different conditions. The need for this kind of information exists, of course, whether or not one uses Bayesian analysis. Judgments about the accuracy and predictability of research findings are implicit in most decisions on whether to order research, regardless of how the decisions are made.

PROCEDURE FOR REVISING PRIOR PROBABILITIES. Alternative procedures for revising prior probabilities on the basis of new information were described in Chapter 5. Little is known about which procedure will give best results under different circumstances. Reports from the field indicate that more use is being made of the procedure of directly assigning revised probabilities than of the Bayesian formula. Although either procedure can be satisfactory, the former would appear to carry more danger that the decision maker might be unduly influenced by the research findings because it does not directly remind him that the research results can be misleading.

POSTERIOR ANALYSIS. Although this section of the chapter has focused on preposterior analysis, most of the comments are applicable to posterior analysis as well.

SEQUENTIAL DECISION MAKING. Much of the attention given to the Bayesian approach has been on its use in the static sense of doing things once to make a choice in a given situation. By plan, however, the approach can become a dynamic process of sequential decision making in which prior probabilities are revised at intervals to reflect changed conditions or new information.

An application of this kind in research design is the use of sequential sampling in which sample size is not fixed in advance. Instead, after a set of observations are made, the new data are analyzed to determine whether still more observations should be made or whether the sampling should be stopped in favor of making a terminal decision on the basis of the information at hand. Sequential sampling frequently requires fewer observations than fixed-size sampling to achieve a given level of reliability.[9]

DOES BAYESIAN ANALYSIS PRODUCE BETTER DECISIONS?

No empirical evidence is available which goes very far toward proving that any given approach to decision under uncertainty, whether formal or informal, results in better decisions than would have been made by some other means. Nor can such evidence be expected considering the

[9] For a more detailed treatment, see Paul E. Green and Donald S. Tull, *Research for Marketing Decisions* (Englewood Cliffs, N.J.: Prentice-Hall, Inc., 1966), pp. 274–283.

difficulty and the size of the research effort that would be entailed. One can only hope for pieces of limited evidence on whether decision ability seems to be enhanced or the decision process facilitated in some way.

Information of any kind on the effects of use of Bayesian analysis is very scarce. Business applications to date, although growing, have been limited in number and subjected to little research. The literature tells of no controlled experiments in which the approach has been compared with another on measures of outcome or observed effects on the decision process. One relevant study, however, has been reported. It was made by Root, who observed applications of a computer simulation model and subjective probability estimates in new product investment decision analyses in four different firms.[10] He found that use of the model increased the collective thinking that went into new product proposals and encouraged more thorough analysis. He noted that subjective probability estimates forced the participants to consider elements of risk they previously had neglected when using single point estimates, and he reported an increase in sensitivity-analysis activity during his period of observation. He also observed that use of the approach did not result in an increase in the number of alternatives which the analysis group developed and submitted for consideration to the decision group of executives.

Some of the users of Bayesian analysis in business have reported informally on their experiences. Reactions known to the author have been somewhat mixed for different reasons, although it is clear that the approach has a growing number of advocates who are convinced that its use has been beneficial to them personally or to the decision-making activity of their organizations.

At the present time, a decision to adopt the Bayesian approach must rest on a judgment that its qualities will benefit the decision process and, perhaps, result in better decisions than would otherwise be made. It should be recognized that the value of the approach may well be different for different decision problems, organizational contexts, and individuals. Although the decision-making ability of many people may be helped if they are forced to be more structured and specific in their thinking, this may not be true for others.

NEED FOR RESEARCH

As the preceding discussion has suggested, there is a need and opportunity for research to develop better answers to a number of questions relevant to effective implementation of Bayesian analysis. The questions differ in their emphasis. Some are largely procedural. For example:

[10] H. Paul Root, "The Analysis of New Products: A Comparative Study of the Evaluation of Product Innovations," (Ph.D. diss., Purdue University, 1969).

How can accurate subjective probability assessments be obtained?

Which procedure for revising prior probabilities on the basis of new information gives best results (for example, one involving the use of the Bayesian formula or direct assignment of the revised probabilities)?

What procedure should be followed in the evaluation and use of the heterogeneous information that results from a research effort consisting of different kinds of inquiries (market tests, product-use tests, surveys, and so forth) in revising prior probabilities?

How can attitude toward risk best be brought to bear in the decision analysis?

Other questions focus on the need for somewhat different kinds of operational information. For example:

For what kinds of decisions and circumstances is the use of the Bayesian approach best suited?

What guidelines can be helpful for determining how detailed a given analysis should be?

How can the time requirements of modeling a complex decision problem be reduced?

What degree of predictability is associated with the different kinds of research observations made under given conditions?

Other questions put more emphasis on organizational and administrative considerations. Here are several.

What organizational and administrative provisions should be made to promote effective implementation of decision analysis?

- To what extent should the responsible decision-making executive be involved?
- Should there be a group of specialists who assume much of the burden of structuring the problem, obtaining the needed inputs, and conducting the analysis?
- What should be the nature of the working relationships between line executives, decision-analysis specialists, and other research personnel in the organization?
- How can executives best be introduced to decision analysis so that they can make effective use of it?

How can the Bayesian approach be applied effectively when decision-making responsibility rests with a group rather than with a single individual?

Although some of these questions are unique to Bayesian analysis, many are not. Problems of structuring, obtaining good judgments and other kinds of inputs, evaluating research findings, and making administrative provisions to facilitate the decision process in an organization, for example, are present regardless of what approach is taken in decision making under uncertainty—as is the need for better answers. The rather formidable list of questions presented in the context of a discussion of the Bayesian approach, therefore, need not deter one from giving it a try. The approach is well developed conceptually, and the technology is far ahead of the current capacity of most business organizations to make use of it. The latter deficiency can be corrected over time by education, experience, and research.

THE OUTLOOK

The use of decision theory by management appears to be in the very early stage of a growth cycle. In a survey conducted as part of a project concerned with some of the questions mentioned here, Brown found a substantial amount of applied work in a handful of companies.[11] One corporation, for example, had conducted 500 decision analyses in the last four years. The several pioneers in the field have had about ten years of experience with the approach. The number of users has expanded rapidly in the past three or four years. Brown reported that several large companies have sponsored orientation courses in decision analysis that have been attended by hundreds of their executives.

Management applications of decision theory can be expected to increase markedly in the future for several reasons. The approach appears to offer a number of potential advantages. Decision analysis is now being taught in leading schools of business, and it has been a subject for attention in a number of management-development programs. Another important reason is the availability of the computer, with its tremendous capacity for computation. Its very presence represents not only opportunity but also pressure for developing applications of various quantitative methods.

Just how much impact the use of decision theory will have on management remains to be seen. There is reason to expect, however, that it will substantially affect the way managers think and act in the process of making decisions under uncertainty. Perhaps the most significant benefit of the use of decision theory in the longer run will come from its focusing attention on the limitations of current knowledge for making the judgments normally demanded of managers. Although the decision model and the computer can be valuable assets, there is no magic in the machinery that will turn bad judgments into good decisions. Greater recognition of this fact should promote systematic inquiry and improved recording and analysis of experiential data to better serve management's decision needs.

[11] Rex V. Brown, "Do Managers Find Decision Theory Useful?" *Harvard Business Review* Vol. 48, no. 3 (May–June 1970):78–89.

APPENDIX

BACKGROUND ON BAYESIAN ANALYSIS OF BUSINESS DECISIONS

Work in applying decision theory in business lagged substantially behind the theoretical development. This condition existed despite advocacy by Savage[1] and other statisticians of the use of subjective probabilities and despite growing interest in the business world in the making of choices under uncertainty stimulated by the presentation in 1959 by Schlaifer[2] of a normative theory addressed to the practitioner as well as to the theorist.

Since the appearance of Schlaifer's book, a number of writings have appeared that have described the Bayesian approach and suggested business applications. Raiffa and Schlaifer presented an introduction to the mathematical analysis of decision making under uncertainty when experimentation is possible.[3] Pratt, Raiffa, and Schlaifer wrote *Introduction to Decision Under Uncertainty*,[4] and they and others at the Harvard Business School prepared other teaching materials on the subject.[5,6] I benefited greatly from seeing parts of a manuscript on making choices under uncertainty while it was being prepared by Raiffa[7] and from attending his lectures on the subject.

Among early descriptions of Bayesian analysis in the business literature were articles by Roberts[8] and Hirshleifer[9] and the marketing-oriented expositions by Roberts,[10] Alderson and Green,[11] Bass,[12] and Green.[13] These writings focused on Bayesian concepts and procedures and suggested potential business applications. Reports of developmental work on applications to actual business problems have been fewer in number, but several appeared in the early 1960s. Grayson[14] and Kaufman[15] gave attention to capital budgeting and decision theory in the context of oil exploration, and Christenson analyzed decision problems encountered by investment bankers in bidding for corporate debt securities.[16] Magee dealt with applications in capital budgeting and market planning,[17] and Hertz described applications of "risk analysis" in investment planning.[18]

In the area of marketing, Green presented a paper, "An Application of Bayesian Decision Theory to a Problem in Long Range Pricing Strategy," at the annual meeting of the American Statistical Association in December, 1961. The application later was described by Green in a journal article[19] and presented in case form by Buzzell.[20] Green also authored other articles dealing with disguised applications of the Bayesian approach.[21] Buzzell and Slater outlined a Bayesian analysis of a policy decision on private branding in the bakery industry.[22] The references cited in this paragraph represent the young state of applied work and the context in which I began my project, which embraced observing and describing selected decision situations in the field and the first phase of work on applying Bayesian decision theory to those situations.[23]

[1] L. J. Savage, *The Foundations of Statistics* (New York: John Wiley & Sons, Inc., 1954).

[2] Robert Schlaifer, *Probability and Statistics for Business Decisions* (New York: McGraw-Hill Book Company, 1959).

[3] Howard Raiffa and Robert Schlaifer, *Applied Statistical Decision Theory* (Boston: Division of Research, Graduate School of Business Administration, Harvard University, 1961).

[4] John W. Pratt, Howard Raiffa, and Robert Schlaifer, *Introduction to Decision Under Uncertainty* (Boston: Graduate School of Business Administration, Harvard University, 1964).

[5] John W. Pratt, Howard Raiffa, and Robert Schlaifer, *Introduction to Statistical Decision Theory* (New York: McGraw-Hill Book Company, 1965).

[6] Robert Schlaifer, *Analysis of Decisions Under Uncertainty* (New York: McGraw-Hill Book Company, 1969).

[7] Howard Raiffa, *Decision Analysis: Introductory Lectures on Choices under Uncertainty* (Reading, Mass.: Addison-Wesley Publishing Co., Inc., 1968).

[8] H. V. Roberts, "The New Business Statistics," *Journal of Business* 33, no. 1 (January 1960): 21–30.

[9] J. Hirshleifer, "The Bayesian Approach to Statistical Decision: An Exposition," *Journal of Business* 34, no. 4 (October 1961): 471–489.

[10] H. V. Roberts, "Bayesian Statistics in Marketing," *Journal of Marketing* 27, no. 1 (January 1963): 1–4.

[11] Wroe Alderson and Paul E. Green, *Planning and Problem Solving in Marketing* (Homewood, Ill.: Richard D. Irwin, Inc., 1964), pp. 82–141.

[12] Frank M. Bass, "Marketing Research Expenditures: A Decision Model," *Journal of Business* 36, no. 1 (January 1963): 77–90.

[13] Paul E. Green, "Bayesian Statistics and Product Decisions," *Business Horizons* 5, no. 3 (Fall 1962): 101–109; idem, "Bayesian Decision Theory in Advertising," *Journal*

of Advertising Research 2, no. 4 (December 1962): 33–41; idem, "The Computer's Place in Business Planning: A Bayesian Approach," in *Marketing and the Computer,* ed. Wroe Alderson and S. J. Shapiro (Englewood Cliffs, N.J.: Prentice-Hall, Inc., 1963) pp. 278–300; idem, "Decision Theory in Market Planning and Research," in *Models, Measurement and Marketing,* ed. Peter Langhoff (Englewood Cliffs, N.J.: Prentice-Hall, Inc., 1965), pp. 171–197.

[14] C. Jackson Grayson, Jr., *Decisions Under Uncertainty—Drilling Decisions by Oil and Gas Operators* (Boston: Division of Research, Graduate School of Business Administration, Harvard University, 1960).

[15] Gordon M. Kaufman, *Statistical Decision and Related Techniques in Oil and Gas Exploration* (Englewood Cliffs, N.J.: Prentice-Hall, Inc., 1963).

[16] Charles Christenson, *Strategic Aspects of Competitive Bidding for Corporate Securities* (Boston: Division of Research, Graduate School of Business Administration, Harvard University, 1965).

[17] John F. Magee, "Decision Trees for Decision Making," *Harvard Business Review* 42, no. 4 (July–August 1964): 126–138; idem, "How to Use Decision Trees in Capital Investment," *Harvard Business Review* 42, no. 5 (September–October 1964): 79–96.

[18] D. B. Hertz, "Risk Analysis in Capital Investment," *Harvard Business Review* 42, no. 1 (January–February 1964): 95–106.

[19] Paul E. Green, "Bayesian Decision Theory in Pricing Strategy," *Journal of Marketing* 27, no. 1 (January 1963): 5–14.

[20] Robert D. Buzzell, "Everclear Plastics Company: Decision Theory Analysis of a Pricing Problem," *Mathematical Models and Marketing Management* (Boston: Division of Research, Graduate School of Business Administration, Harvard University, 1964), pp. 112–135.

[21] Paul E. Green, "Decision Theory and Chemical Marketing," *Industrial and Engineering Chemistry* 54 (September 1962): 30–34; idem, "Decisions Involving High Risk," *Advanced Management—Office Executive* 1 (October 1962): 18–23.

[22] Robert D. Buzzell and Charles C. Slater, "Decision Theory and Marketing Management," *Journal of Marketing* 26, no. 3 (July 1962): 7–16.

[23] Joseph W. Newman, "An Application of Decision Theory Under the Operating Pressures of Marketing Management," Working Paper No. 69 (Graduate School of Business, Stanford, Calif. Stanford University, August 1965).

INDEX

74194

71 72 73 74 7 6 5 4 3 2 1